Musculoskeletal Examination
——————— of the ———————
Shoulder

Making the Complex Simple

MUSCULOSKELETAL EXAMINATION
MAKING THE COMPLEX SIMPLE
SERIES

Editor, Steven B. Cohen, MD

Musculoskeletal Examination
——— of the ———
Shoulder
Making the Complex Simple

Editor:

Steven B. Cohen, MD
Rothman Institute
Department of Orthopaedics
Sports Medicine
Philadelphia, Pennsylvania

SLACK
INCORPORATED

www.slackbooks.com

ISBN: 978-1-55642-912-5

Copyright © 2011 by SLACK Incorporated

The procedures and practices described in this book should be implemented in a manner consistent with the professional standards set for the circumstances that apply in each specific situation. Every effort has been made to confirm the accuracy of the information presented and to correctly relate generally accepted practices. The authors, editor, and publisher cannot accept responsibility for errors or exclusions or for the outcome of the material presented herein. There is no expressed or implied warranty of this book or information imparted by it. Care has been taken to ensure that drug selection and dosages are in accordance with currently accepted/recommended practice. Due to continuing research, changes in government policy and regulations, and various effects of drug reactions and interactions, it is recommended that the reader carefully review all materials and literature provided for each drug, especially those that are new or not frequently used. Any review or mention of specific companies or products is not intended as an endorsement by the author or publisher.

SLACK Incorporated uses a review process to evaluate submitted material. Prior to publication, educators or clinicians provide important feedback on the content that we publish. We welcome feedback on this work.

Published by: SLACK Incorporated
                     6900 Grove Road
                     Thorofare, NJ 08086 USA
                     Telephone: 856-848-1000
                     Fax: 856-848-6091
                     www.slackbooks.com

Contact SLACK Incorporated for more information about other books in this field or about the availability of our books from distributors outside the United States.

Library of Congress Cataloging-in-Publication Data

Musculoskeletal examination of the shoulder : making the complex simple / [edited by] Steven B. Cohen.
     p. ; cm.
Includes bibliographical references and index.
ISBN 978-1-55642-912-5 (alk. paper)
1. Shoulder--Examination--Handbooks, manuals, etc. 2. Shoulder--Diseases--Diagnosis--Handbooks, manuals, etc. I. Cohen, Steven B., M.D.
[DNLM: 1. Physical Examination--methods--Handbooks. 2. Shoulder--injuries--Handbooks. 3. Musculoskeletal Diseases--diagnosis--Handbooks. WE 39]
RC939.M87 2011
617.5'72--dc22
                              2010026774

Printed in the United States of America.

Last digit is print number: 10 9 8 7 6 5 4 3 2 1

DEDICATION

To Kathleen, Alexa, Will, and Ty. Without all of you, none of this would be possible. You give me motivation and purpose, and I am so thankful for having such a beautiful family.

CONTENTS

ACKNOWLEDGMENTS

I would like to acknowledge Carrie Kotlar, John Bond, and the rest of the staff at SLACK Incorporated. Thank you for seeing this project to completion and for your trust and confidence in me.

ABOUT THE EDITOR

Steven B. Cohen, MD grew up in Cherry Hill, New Jersey and attended Columbia University in New York, New York, where he was a pre-medicine and political science major while playing varsity baseball. He received his medical degree from the University of Medicine and Dentistry of New Jersey—Robert Wood Johnson Medical School in Piscataway, New Jersey. Following graduation, he completed his orthopedic surgery residency at the University of Virginia Health Sciences Center in Charlottesville, Virginia. He then completed a sports medicine fellowship at the University of Pittsburgh Medical Center, Pittsburgh, Pennsylvania.

He is a sports medicine specialist at the Rothman Institute in Philadelphia, Pennsylvania and serves as the Director of Sports Medicine Research. He is currently an assistant professor in the Department of Orthopedic Surgery at Thomas Jefferson University in Philadelphia, Pennsylvania. He is the Assistant Team Physician for the Philadelphia Phillies and St. Joseph's University and the Medical Director for the Philadelphia Marathon. He serves on several committees for the American Orthopedic Society for Sports Medicine and International Society of Arthroscopy, Knee Surgery, and Sports Medicine. He has been the editor of *Current Concepts in ACL Reconstruction* with Dr. Freddie H. Fu, the guest editor of *Clinics in Sports Medicine*, and the associate editor of the *Textbook of Arthroscopy* with Drs. Mark Miller and Brian Cole. He has authored over 110 peer-reviewed manuscripts, book chapters, posters, and national and international presentations.

He currently lives with his wife, Kathleen, and 3 children, Alexa, Will, and Ty, in Media, Pennsylvania.

CONTRIBUTING AUTHORS

Geoffrey S. Baer, MD, PhD (Chapter 7)
Assistant Professor, Orthopedic Surgery
Department of Orthopedics and Rehabilitation
Division of Sports Medicine
University of Wisconsin—Madison
Madison, Wisconsin

Michael G. Ciccotti, MD (Chapter 4)
Professor, Orthopedics
Chief, Division of Sports Medicine
Director, Sports Medicine Fellowship
Department of Orthopaedics
Rothman Institute
Thomas Jefferson University
Head Team Physician, Philadelphia Phillies Baseball and St.
   Joseph's University
Philadelphia, Pennsylvania

David R. Diduch, MD (Chapter 5)
Alfred R. Shands Professor, Orthopedic Surgery
Head Orthopedic Team Physician
Fellowship Director, Sports Medicine
Department of Orthopaedic Surgery
University of Virginia
Charlottesville, Virginia

Charles L. Getz, MD (Chapter 9)
Assistant Professor, Orthopedic Surgery
Thomas Jefferson Medical School
Rothman Institute
Thomas Jefferson University Hospital
Philadelphia, Pennsylvania

George Paul Hobbs, MD (Chapter 2)
Department of Musculoskeletal Radiology
Thomas Jefferson University
Philadelphia, Pennsylvania

Gregg J. Jarit, MD (Chapter 5)
Orthopedic Associates of Long Island
East Setauket, New York

Scott Montgomery, MD (Chapter 1)
Ochsner Medical Center
New Orleans, Lousiana

William B. Morrison, MD (Chapter 2)
Professor, Radiology
Director, Division of Musculoskeletal Radiology
Thomas Jefferson University Hospital
Philadelphia, Pennsylvania

Mark W. Rodosky, MD (Chapter 6)
Assistant Professor, Orthopedic Surgery
Division of Sports Medicine
Chief, Shoulder Service
UPMC Center for Sports Medicine
Pittsburgh, Pennsylvania

James R. Romanowski, MD (Chapter 6)
Fellow, Orthopedic Sports Medicine
University of Pittsburgh School of Medicine
UPMC Center for Sports Medicine
Pittsburgh, Pennsylvania

Michael Shin, MD (Chapter 7)
Orthopedic Surgeon
Valley Orthopedics and Sports Medicine
Saint Margaret's Hospital
Spring Valley, Illinois

Misty Suri, MD (Chapter 1)
Ochsner Medical Center
New Orleans, Lousiana

Michael R. Tracy, MD (Chapter 9)
Clinical Instructor in Surgery (Orthopaedics)
The Commonwealth Medical College
Scranton, Pennsylvania
Shoulder and Elbow Surgeon
Scranton Orthopaedic Specialists
Dickson City, Pennsylvania

Gerald R. Williams Jr, MD (Chapter 8)
Professor and Chief, Shoulder and Elbow Service
Rothman Institute at Jefferson
Jefferson Medical College
Philadelphia, Pennsylvania

FOREWORD

The trend in medicine is to specialize and concentrate expertise in one area. The *Musculoskeletal Examination* book series focuses specifically on the prevention, diagnosis, treatment, and rehabilitation of injuries due to sport, exercise, recreational activity, or trauma. The series of 5 books edited by Steve Cohen is an exceptional resource for not only the specialist but for any clinician at all levels of training.

Musculoskeletal Examination of the Shoulder: Making the Complex Simple is important on several levels. This text is valuable as both an educational tool as well as a ready reference. In addition, the book is an essential resource for the practitioner who is not a specialist, but finds himself or herself caring for individuals with all types of shoulder injuries.

As a member of the faculty, I worked with Steve Cohen as he completed his sports medicine fellowship at the University of Pittsburgh Medical Center. Since that time, Steve has gone on to excel in his field as a sports medicine orthopedic specialist and in working with professional athletes. He is a respected colleague and published author. I commend Steve Cohen for his efforts to advance the practice of medicine through this salient publication.

James P. Bradley, MD
Clinical Professor, Orthopedics
University of Pittsburgh Medical Center
Team Physician, Pittsburgh Steelers
Pittsburgh, Pennsylvania

INTRODUCTION

Treatment of shoulder disorders dates back to Hippocrates' original description of reducing shoulder dislocations. Shoulder problems are common in all age groups regardless of activity level. From the young, active patient with growth plate injuries to the mature adult with rotator cuff pathology, treatment of the shoulder is based largely on a thorough examination of the shoulder.

Specific shoulder injuries can frequently be grouped and determined by age. The younger population (<age 40) more often presents with specific traumatic injuries to the shoulder, which include instability, labral tears, and acromioclavicular joint separation. The more middle-aged population (age 40 to 65) may present with overuse injuries to the rotator cuff, biceps tendon, and acromioclavicular joint arthritis. While the older population (>60 years old) can also present with rotator cuff pathology, degenerative arthritis of the glenohumeral joint is also common. Traumatic fracture about the shoulder can occur in any age group but is more frequently seen in the older population.

Treatment of these disorders varies based on the pathology. In general, nonsurgical treatment is recommended except in the cases of acute injuries such as fractures and rotator cuff tears. In a large majority of cases, rest and rehabilitation allow a return to function. Surgical management can include arthroscopy or open surgery depending on the pathology and surgeon preference.

The goal of this book is to provide those who evaluate and treat conditions of the shoulder, including instability, labral tears, rotator cuff tendonitis/tears, biceps tendon disorders, acromioclavicular joint problems, fractures, and arthritis, with the most up-to-date information. In particular, as the title describes, examination of the shoulder is thoroughly reviewed. Additionally, there is a chapter on imaging of the shoulder from the simple x-ray to the more complex magnetic resonance imaging. The authors of each chapter in this book are well-respected, experienced shoulder surgeons who have contributed to the literature on the diagnosis and treatment of shoulder disorders. Each author was instructed to keep the information within a simple outline and to avoid spending paragraphs on uncommon or esoteric problems. The information in each chapter should be useful as a guide to the physical diagnosis of most shoulder conditions. For more detailed information or for greater expertise in an area of shoulder surgery, this book will point the way toward advanced diagnosis and treatment and explain the need for experienced shoulder surgeons in caring for the difficult and complex clinical issues.

I

Physical Examination

1

PHYSICAL EXAMINATION OF THE SHOULDER

THE BASICS AND SPECIFIC TESTS

Scott Montgomery, MD and Misty Suri, MD

INTRODUCTION

After an accurate history, more information is obtained from physical examination than from any other source. Before expensive imaging is used to look deep within the patient's shoulder joint, a complaint-focused physical exam can provide useful information. Though some may argue that magnetic resonance imaging (MRI) can tell much more than any other source of information, an astute clinician can and should extract specific useful information from physical examination of the shoulder joint. In that light, it is always most important to treat the patient and not an imaging study.

Cohen SB. *Musculoskeletal Examination of the Shoulder:*
Making the Complex Simple (pp. 2-22).
© 2011 SLACK Incorporated

INSPECTION/SURFACE ANATOMY

The physician can gather a significant amount of information from the patient by simply observing. Having the patient in a gown tied around the chest below the axilla is a good technique to allow visualization of both shoulders, including both scapulae. This allows the female patient to be appropriately covered with full shoulder exposure. Inspect the skin, noticing any surgical scars, deformities or asymmetries. Specifically, on the anterior side of the body, visually examine both clavicles: the sternoclavicular (SC) joints and the acromioclavicular (AC) joints. Notice whether the SC joints are recessed or prominent. Determine whether either distal clavicle is elevated or more prominent at the AC joint than the other. Also, notice any prominence along the shaft of the clavicle. The coracoid is the anterior most part of the scapula and is a small bony prominence inferior and medial to the AC joint. The acromion is the lateral aspect of the scapula and is part of the AC joint as well as the origin of the deltoid muscle. The posterolateral corner of the acromion is an important landmark in planning treatments such as subacromial injections or starting arthroscopy portals.

A shoulder examination includes the anterior aspect of the chest wall. The pectoralis muscle extends from the anterior chest wall to the proximal humerus. The pectoralis major muscle forms the anterior aspect of the axilla. *Rupture of the pectoralis major* muscle results in loss of the axillary webbing or loss of the anterior axillary fold.

The deltoid muscle should be examined at its origin, muscle belly, and insertion to ensure its competence and symmetry. *Deltoid dehiscence* is a loss of continuity off of the acromion of one or more of the heads of the deltoid and can result in a retraction or balling-up of the muscle belly. This can happen traumatically or, more commonly, as a result of a postoperative complication. *Deltoid atony* is the loss of muscle innervation or muscle tone to the deltoid, without a discontinuity of its origin. The result is the appearance of a sulcus sign, or inferior subluxation of the glenohumeral joint. The patient appears to have a "droopy shoulder" along with a decrease in the muscle tone of the deltoid itself. This clinical scenario can result from

an axillary nerve injury, a loss of deltoid tone post injury, or a
deconditioning of the deltoid muscle belly itself.
The biceps muscle belly should be included in any inspection of the shoulder joint. Symmetry and size are noted.
Particular attention is paid to whether a dropped biceps or
"Popeye" muscle deformity is present, indicating a rupture of
the proximal long head of the biceps brachii.
Posteriorly, the shoulder inspection begins at the midline
and proximally along the lateral side of the neck. The trapezius
muscle originates along the spinous processes of both the cervical and thoracic vertebrae. This muscle inserts on the spine
of the scapula and acromion. It is visualized for signs of asymmetry, possibly indicating a spinal accessory nerve injury.
The scapula is visualized from the posterior side. The static
position of the scapula is noted with the arm at the side. The
affected scapula is compared with the other side by measuring
the inferior position, the lateral displacement, and the abduction.[1] The spine of the scapula divides the supraspinatus fossa
from the infraspinatus fossa. Asymmetry is usually a sign of
muscle atrophy of either of the rotator cuff muscle origins.
Atrophy of the supraspinatus fossa alone can indicate injury
or impingement of the suprascapular nerve at the suprascapular notch. Muscle atrophy in both the supraspinatus and
infraspinatus fossae can indicate injury or impingement of the
suprascapular nerve at the spinoglenoid notch.

PALPATION

After visual inspection, the shoulder is palpated. If one side
is more symptomatic, start with the unaffected side to estab-
lish a sense of each patient's "normal" as well as the patient's
trust. There is a great deal of variability from patient to patient
and the importance of having a good understanding of what a
patient's baseline examination is cannot be overstated. Muscle
tone is assessed for symmetry, specifically palpating the deltoid, biceps, triceps, pectoralis, and trapezius as discussed
above.
The specific location of pain is important. Ask the patient to
point with one finger using the opposite hand. Common locations will include the top of the AC joint, anteriorly along the

proximal biceps tendon, laterally along the deltoid insertion, or in the back along the posterolateral corner of the acromion. Often, the patient will use his or her entire palm to rub over the lateral aspect of the upper arm or will say that the pain is "deep" in his or her shoulder and will be unable to target the pain more specifically. Be careful to elicit any signs of cervical radiculopathy. Frequently, patients who complain of "shoulder pain" will in fact have pain that radiates down the entire arm. It is important to specifically ask if any pain, numbness, or tingling occurs in the patient's hand and fingertips. Symptoms that travel distal to the elbow more commonly originate from the neck than the shoulder. If the patient remains vague and complains of both shoulder/upper arm and radiating pain, ask which bothers him or her the most and explain why the difference is important. Concurrent problems with the shoulder and cervical spine can and do occur, but first try to focus the patient and yourself to looking for a single cause for his or her symptoms.

Points of tenderness are then palpated. Distracting techniques may be employed to ensure that the specific points of tenderness are accurate and reproducible. Direct palpation can be uncomfortable on a normal AC joint or proximal biceps tendon sheath, so ensure that the unaffected side is checked and compared. The patient should be able to tell you that the affected side is painful or more symptomatic than normal. Pertinent positives may be quickly rechecked after another test or 2 are performed to ensure the tender points are valid and reproducible and confirmed with the patient that they are an important part of his or her symptoms. Patients being examined in clinics associated with training programs can often change their story after the initial test when they get a chance to truly think about their symptoms and what bothers them the most. Attending physicians might then get a different story when the joint is re-examined. Nonacademic or solo practitioners can ensure the patient gets a chance to think or regroup by simply rechecking positives after several other tests have been performed.

Proximal biceps tendon sheath tenderness can be a primary or secondary finding. Inflammation in the proximal biceps tendon sheath, or tenosynovitis, with an intact tendon can produce tenderness. Primary biceps tendonopathy with a degenerative or partially torn tendon in the bicipital groove can also be tender. These can often be seen in conjunction with the patient who has rotator cuff pathology or symptoms of external outlet impingement. Secondary biceps tenosynovitis and proximal tenderness can be seen in younger patients with underlying shoulder instability and should not be confused as the primary underlying problem.

The acromion forms from several underlying growth plates. An os acromiale is an acromion that fails to fuse between 2 of the underlying physes. The most common area is in between the meso- and meta-acromion. This is often incidental, asymptomatic, and bilateral. If identified on axillary lateral x-ray or MRI, the os acromiale is palpated to illicit any tenderness. If tender and verified as a source of the patient's complaints, diagnostic injection with lidocaine/marcaine directly into the unfused physes can be performed. If the symptoms are alleviated and the os acromiale is found to be unstable on arthroscopy, consideration may be given to surgical fusion or removal of the unstable os acromiale.

RANGE OF MOTION

The shoulder has more freedom of motion than any other joint in the body. It allows us to place our arms above our heads and reach out in front of and behind our backs. Normal shoulder motion gains its range of motion (ROM) through both the glenohumeral joint and the scapulothoracic joint in a 2:1 ratio. Average shoulder ROM varies from patient to patient but is estimated below[2]:

- Abduction: 180 degrees
- Horizontal adduction: 30 to 45 degrees
- Flexion: 180 degrees
- Extension: 45 degrees
- Internal rotation: 55 degrees
- External rotation: 45 degrees

The posterior aspect of the scapula should be viewed during shoulder motion. Motion between the posterior chest wall and the anterior scapula is called *scapulothoracic motion*. This can be limited from trauma, scarring or adhesions, tumor, or arthritis of the SC or AC joint.

Scapular dyskinesia is any abnormal movement disorder of the scapula and is most easily noted by comparing the motion of the affected scapula versus the unaffected side. *Medial scapular winging* can be a result of a long thoracic nerve palsy and subsequent serratus anterior weakness. *Lateral scapular winging* can be the result of trapezius muscle palsy from injury to the spinal accessory nerve. These entities can be difficult to distinguish clinically. A dynamic exam of the scapula is performed with repeated full forward elevation of both arms while observing from behind. Any asynchronous or dyskinetic scapular motion can be assessed in this way. Resistance to forward elevation at 30 degrees can accentuate scapular motion abnormalities. The important concept is to look at the motion of the scapula and to recognize if it is normal or not. If it is not, further work-up or testing may be indicated.

Glenohumeral joint motion should be examined both passively (PROM) and actively (AROM). AROM is typically checked first. The patient is examined in the standing position. Bilateral forward flexion, abduction in the plane of the scapula, external rotation, and internal rotation both with the arm adducted to the side and abducted to 90 degrees are checked initially. Decreased AROM can be for a number of reasons, including effort, pain and guarding, weakness from rotator cuff tears or tendonopathy, subacromial bursitis, degenerative arthritis, or capsular contracture.

If AROM is decreased, PROM is then checked. The patient is asked to sit on the examination table. The physician stabilizes the scapula with one hand and passively moves the shoulder through the same motion planes. It is important to get the patient to relax during this portion of the examination. Patients are often reluctant to allow PROM of a painful joint by the examiner. A good pearl is to gently grasp the elbow and internally and externally rotate the arm through the shoulder as you stabilize the scapula. This often allows the patient to drop the weight of his or her arm into your hand and allows you to examine it. The examiner can gently stress the passive motion at the point where the active motion stopped. Decreased AROM and PROM can indicate adhesive capsulitis or degenerative arthritis. Increased external rotation at

0 degrees of abduction can be indicative of a subscapularis tear. Radiographic studies are used to confirm arthritis. Normal or minimal changes on radiographs with decreased PROM and AROM indicate a stiff shoulder. Shoulder stiffness is often lumped together but can be from various causes and has specific names reflecting its etiology. The most common are post-traumatic stiffness, postoperative stiffness, diabetic frozen shoulder, or idiopathic adhesive capsulitis. The final common pathway is decreased motion, and treatment should be focused at improving motion. The method to achieve improved motion depends on whether the stiffness is primarily capsular (diabetic frozen shouler, adhesive capsulitis) or involves multiple tissue planes (post-traumatic and postoperative stiffness).[3]

STRENGTH TESTING

The rotator cuff is an important common insertion on the humerus of 4 muscles originating from the scapula. These 4 muscles can be isolated and should be assessed during every shoulder examination. Muscle strength is graded as follows[4]:

- Grade 5—Normal strength
- Grade 4—Near normal strength
- Grade 3—Movement against gravity
- Grade 2—Movement with gravity eliminated
- Grade 1—Visible contraction with no movement
- Grade 0— No palpable contraction

The *supraspinatus* originates from the supraspinatus fossa of the scapula and inserts on the greater tuberosity of the humerus. It is isolated by abduction of the arm 90 degrees in the plane of the scapula (30 degrees horizontally adducted). The arm is then internally rotated so the patient's thumbs face the floor. Resisting elevation (encouraging the patient to raise arms to the ceiling) is graded as above. This motion tests *scaption* and is also called the *empty can test* (Figure 1-1).

The *infraspinatus* originates from the infraspinatus fossa of the scapula and inserts on the greater tuberosity of the humerus. It is isolated by external rotation of the arm at the

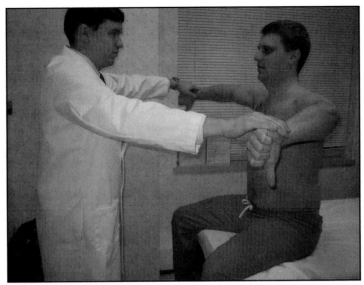

Figure 1-1. Empty can test.

side with the elbow flexed 90 degrees. Resisted external rotation (rotating out) is graded accordingly.

The *subscapularis* originates from the subscapularis fossa on the anterior aspect of the scapula and inserts on the lesser tuberosity of the humerus. The lift off is performed when the patient moves his or her wrist off his or her back against resistance. To do the belly press test, the patient presses his or her hand against his or her abdomen with elbows parallel to the coronal plane of his or her body. A positive belly press test is evident when the patient's wrist flexes or the elbow drops behind his or her body. Subscapularis weakness can be assessed with the belly press test, lift off test, and bear hug test (Figures 1-2 through 1-4).[5-7]

The difference between make testing and break testing to assess muscle strength should be noted. Make testing assesses isometric muscle contraction by resisting the patient's force under static circumstances. Break testing assesses eccentric muscle contraction by overcoming the patient's force until it breaks or gives way. Break testing can be a more reliable test for organic or actual muscle weakness as opposed to functional weakness without known cause.[8,9]

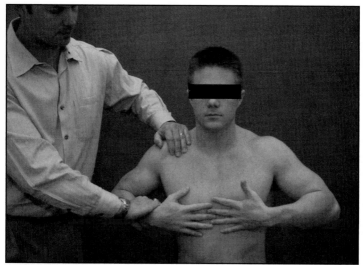

Figure 1-2. Belly press test. (Reprinted with permission from Michael J. Kissenberth, MD.)

Figure 1-3. Lift off test. (Reprinted with permission from Michael J. Kissenberth, MD.)

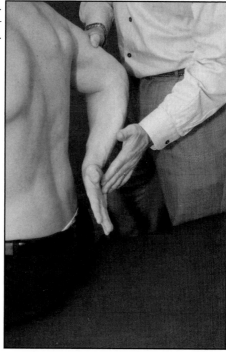

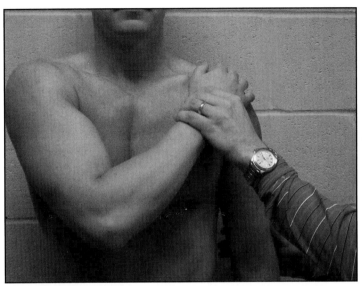

Figure 1-4. Bear hug test.

Rotator cuff weakness can be indicative of a tear in the tendon insertion or muscle belly weakness/atrophy. Atrophy of the muscle, as mentioned previously, can be from impingement, tear, or nerve entrapment. Weakness of both the supraspinatus and infraspinatus can indicate injury or impingement of the suprascapular nerve at the suprascapular notch. Weakness of the infraspinatus alone can indicate injury or impingement of the suprascapular nerve at the spinoglenoid notch (Figure 1-5).

The *deltoid* is best tested by isolating it in thirds. The arm is adducted at the side with the elbow flexed 90 degrees. The anterior third is tested by asking the patient to push away your hand with his or her fist while palpating the deltoid muscles firing with the other hand. The middle third is tested by asking the patient to abduct his or her arm against resistance. The posterior third is tested by asking the patient to extend (push back) the arm against resistance.

The *biceps* is best isolated with the arm adducted at the side with the elbow flexed 90 degrees. Supination against resistance (palm up) and elbow flexion strength are the most effective tests. The *triceps* is isolated by testing resisted elbow extension.

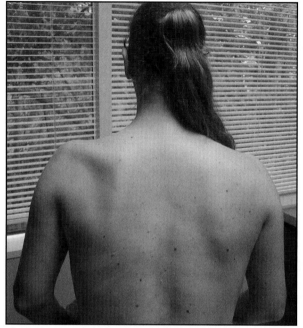

Figure 1-5. Supra-spinatus and infraspinatus atrophy. (Reprinted with permission from Michael J. Kissenberth, MD.)

SENSORY TESTING

The sensation of the entire upper extremity should be tested when examining the shoulder. The *axillary nerve* sensation is best localized on the lateral side of the upper arm. It is best tested by comparing it with the other side for 2-point discrimination. The SC nerve provides sensation to the superior aspect of the shoulder. The lateral cutaneous nerve of the arm provides sensation to the lateral aspect of the arm just distal the axillary nerve distribution.

SPECIFIC TESTS

Impingement Tests

External impingement or outlet impingement occurs when the patient reaches his or her arm above shoulder level causing the undersurface of the acromion to abut the underlying rotator cuff and bursa. Patients often complain of pain in a particular position (overhead). Specific tests have been developed to reproduce those positions.

The Neer impingement sign[10] is performed with the patient seated. The examiner is positioned behind the patient and rests his opposite hand (left hand for right shoulder) on the scapula for stabilization. The examiner maximally forward flexes the patient's shoulder with the arm in internal rotation. A positive test is illicited when pain is felt at maximal forward flexion as the undersurface of the acromion abuts the humerus, impinging on the rotator cuff (see Chapter 5).

The Hawkins impingement sign[11] is also performed with the patient seated. The examiner is positioned to the patient's affected side and rests his or her opposite hand (left hand for right shoulder) on the scapula for stabilization. The shoulder is forward flexed and stabilized at 90 degrees. The examiner grasps the elbow and internally rotates it past 90 degrees until the acromion abuts the humerus. A positive test is pain at the terminal internal rotation.

The impingement test can be verified by a diagnostic injection of lidocaine/marcaine in the subacromial space followed by repeating the Neer and Hawkins tests. A positive impingement test can be verified when the impingement signs resolve and the Neer and Hawkins tests become negative after the injection.

Acromioclavicular Joint Tests

The AC joint can be assessed in several ways (see Chapter 7). Tenderness directly over the AC joint, which is more significant than the opposite side, can indicate arthritis, distal clavicle osteolysis, or traumatic injury such as a fracture or sprain. Pain at the AC joint when the arm is brought across the body is called a positive cross-arm adduction test. When taken together, these tests are suggestive of symptomatic AC joint pathology. Isolated MRI findings of a hypertrophic AC joint without symptoms are most often of little clinical significance. The AC joint can be injected with lidocaine/marcaine and re-examined to determine if the patient's symptoms decrease or completely subside.

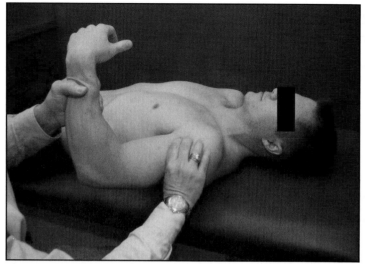

Figure 1-6. Anterior translation testing. (Reprinted with permission from Michael J. Kissenberth, MD.)

Stability Tests

Several tests are available for the clinician to evaluate the shoulder for glenohumeral instability (see Chapter 3). Stability testing requires a cooperative, relaxed patient because even mild guarding can make gathering of valid information difficult. These tests are most easily performed as part of a routine examination under anesthesia prior to surgical intervention.

The load and shift maneuver is performed with the patient supine, testing for directional stability. The arm is held in 90 degrees of abduction and neutral rotation while a force is applied in an attempt to translate the humeral head over the glenoid. A key with this test is to apply a slight axial load to the arm through the elbow to help center the humeral head on the glenoid. Anterior, posterior, and inferior translations are then assessed. Translation is measured as "Grade 1" if the humeral head translates to the glenoid rim (<1 cm); "Grade 2" if the humeral head translates to perch on the glenoid rim (1 to 2 cm); and "Grade 3" if the humeral head translates over the glenoid rim (>3 cm; Figures 1-6 and 1-7).[12,13]

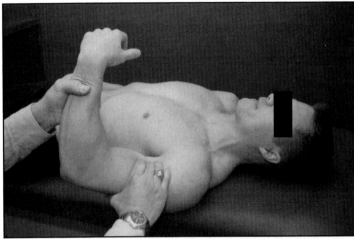

Figure 1-7. Posterior translation testing. (Reprinted with permission from Michael J. Kissenberth, MD.)

A sulcus sign is evaluated in each patient and compared with the contralateral shoulder. The patient is placed in the seated position. The arm is adducted and the scapula is stabilized by the opposite hand while the arm is gently pulled inferiorly. The sulcus is graded 1 through 3 and is tested in neutral and external rotation. A sulcus sign graded as 2 or 3 that does not reduce by one grade when the arm is externally rotated 25 to 30 degrees may indicate an incompetent rotator interval or multidirectional instability. Rotator interval injuries can be seen in isolation but are more frequently part of a combined instability pattern.[14]

The anterior apprehension test is performed with the patient supine. The arm is passively placed in 90 degrees of abduction and the examiner slowly externally rotates the shoulder. The table is used to stabilize the scapula and the patient must be near the edge of the table to allow greater than 90 degrees of rotation. A positive test is when the patient expresses an uncomfortable feeling and feels apprehension that the shoulder will sublux or dislocate as the test reproduces the "at risk" position (Figure 1-8).

The relocation test is performed directly after the apprehension test. With the shoulder in the same position (abducted

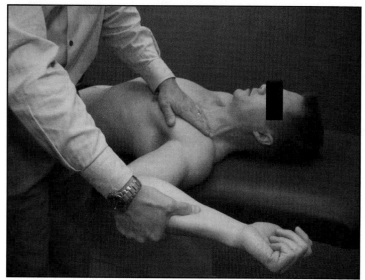

Figure 1-8. Apprehension test. (Reprinted with permission from Michael J. Kissenberth, MD.)

90 degrees, externally rotated 90 degrees or greater) the examiner places the free hand over the anterior humeral head and applies a posteriorly directed force. This action stabilizes the shoulder. The test is positive if the patient feels less apprehensive that the shoulder is going to dislocate (Figure 1-9).[12]

The jerk test is for posterior instability and is performed with the patient in the supine position. The arm is forward flexed 90 degrees and the shoulder is internally rotated while the elbow is flexed at least 90 degrees. The examiner applies a posteriorly directed force through the long axis of the humerus. The examiner is attempting to subluxate or carefully dislocate the shoulder posteriorly over the rim of the glenoid. The arm is then carefully abducted and the humeral head is allowed to reduce or "jerk" back into the glenoid. A positive test is when the patient's symptoms are reproduced, sometimes with an audible or palpable clunk (Figure 1-10).

Biceps/Labral Tests

The active compression test (O'Brien's test[15]) is performed with the patient seated. The examiner is at the side. The

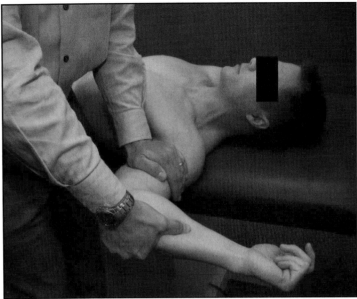

Figure 1-9. Relocation test. (Reprinted with permission from Michael J. Kissenberth, MD.)

patient's shoulder is forward flexed 90 degrees with the elbow in full extension and the arm is placed in 10 degrees of horizontal adduction. The patient resists the examiner's downward force with the arm maximally internally rotated (thumbs down). The test is repeated with the arm maximally externally rotated (palm up). A positive test is more painful in internal rotation than in external rotation (Figure 1-11). Deep pain may indicate indicate a superior labral (SLAP) tear or proximal biceps pathology. O'Brien's test can be positive for AC joint pathology if, while performing the test, the pain is primarily located at the AC joint.

Thoracic Outlet Tests

The Adson test is performed with the patient seated or standing. The examiner palpates the radial pulse with the patient's arm extended and externally rotated. The patient extends and rotates his or her neck toward the affected side and takes a

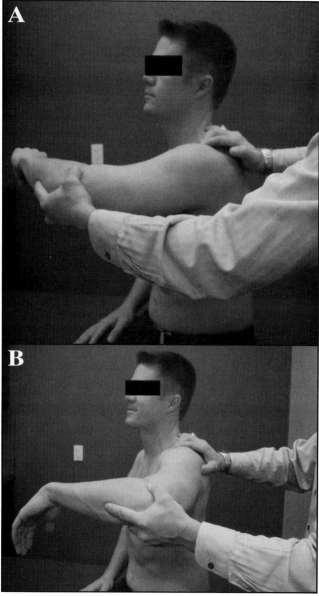

Figure 1-10. Jerk test. (Reprinted with permission from Michael J. Kissenberth, MD.)

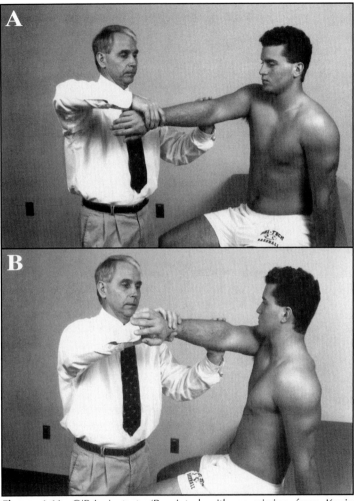

Figure 1-11. O'Brien's test. (Reprinted with permission from Konin JG, Wiksten DL, Isear Jr JA, Brader H. *Special Tests for Orthopedic Examination.* 3d ed. Thorofare, NJ: SLACK Incorporated; 2006.)

deep breath (Figure 1-12). A diminished or absent radial pulse is a positive test and is suggestive of thoracic outlet syndrome from the compression most commonly of the subclavian artery by the scalene muscles or accessory rib. The Roos test is another examination technique for thoracic outlet syndrome.

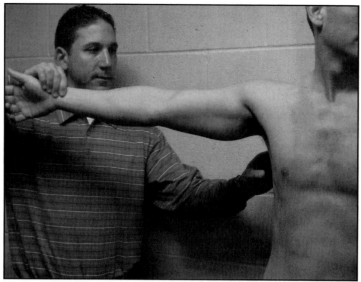

Figure 1-12. Adson test.

Scapular Tests

Scapular dyskinesia tests are indicated if abnormal scapular motion is noted with motion or if scapular symptoms are present. Medial and lateral scapular winging can indicate muscle palsy or nerve injury. A wall push-up is performed by standing several feet away from the wall, leaning into the wall, then pushing away from the wall so the arm goes into a flexed position (Figure 1-13). Moderate or severe cases of scapular winging can be detected when the arm is at the patient's side or with the arm flexed. Mild cases can be detected with a wall push-up. Scapular dyskinesia can be detected with repeated full forward elevation while the examiner observes from behind. Observation of the normal scapular rhythm can be appreciated with this maneuver. A scapular flip sign, as described by Kelly, can be used to help confirm serratus anterior weakness and long-thoracic nerve palsy.[3] With the physician behind the patient, the arm is adducted and the elbow flexed 90 degrees. Resisted external rotation of the shoulder causes the medial border of the affected scapula to become more prominent or "flip," indicating weakness of the serratus anterior. This is

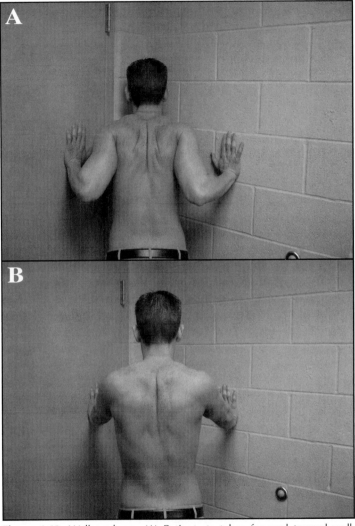

Figure 1-13. Wall push-up. (A) Patient stretches forward toward wall. (B) Patient pushes outward away from wall and scapula is assessed for winging.

judged as positive or negative and is not graded. Winging can can also be elicited with forward elevation resisted at 30 degrees, performing a wall push-up, or by observing the patient using his or her arms to get out of a chair.[16]

REFERENCES

1. Beighton PH, Horan FT. Dominant inheritance in familial generalized articular hypermobility. *J Bone Joint Surg Br.* 1970;52:145-147.
2. Hawkins RJ, Misamore GW. *Shoulder Injuries in the Athlete: Surgical Repair and Rehabilitation.* New York, NY: W.B. Saunders; 1996.
3. Kelley MJ, Kane TE, Leggin BG. Spinal accessory nerve palsy: associated signs and symptoms. *J Orthop Sports Phys Ther.* 2008;38(2):78-86.
4. Krishnan SG, Hawkins RJ, Bokor DJ. Clinical evaluation of shoulder problems. In: Rockwood CA Jr, Masten FA 3rd, eds. *The Shoulder.* Vol 1. 3rd ed. Philadelphia, PA: Saunders; 2004;165.
5. Gerber C, Krushell RJ. Isolated rupture of the tendon of the subscapularis muscle. Clinical features in 16 cases. *J Bone Joint Surg Br.* 1991;73(3):389-394.
6. Tokish JM, Decker MJ, Ellis HB, Torry MR, Hawkins RJ. The belly-press test for the physical examination of the subscapularis muscle: electromyographic validation and comparison to the lift-off test. *J Shoulder Elbow Surg.* 2003;12(5):427-430.
7. Barth JR, Burkhart SS, De Beer JF. The bear-hug test: a new and sensitive test for diagnosing a subscapularis tear. *Arthroscopy.* 2006;22(10):1076-1084.
8. Bang MD, Deyle GD. Comparison of supervised exercise with and without manual physical therapy for patients with shoulder impingement syndrome. *J Orthop Sports Phys Ther.* 2000;30(3):126-137.
9. van der Ploeg RJ, Oosterhuis HJ. The "make/break test" as a diagnostic tool in functional weakness. *J Neurol Neurosurg Psychiatry.* 1991;54(3):248-251.
10. Neer CS 2nd. Impingement lesions. *Clin Orthop Relat Res.* 1983;(173):70-77.
11. Hawkins RJ, Kennedy JC. Impingement syndrome in athletes. *Am J Sports Med.* 1980;8(3):151-158.
12. Silliman JF, Hawkins RJ. Classification and physical diagnosis of instability of the shoulder. *Clin Orthop Relat Res.* 1993;(291):7-19.
13. Ramappa AJ, Hawkins RJ, Suri M. Shoulder disorders in the overhead athlete. *Instr Course Lect.* 2007;56:35-43.
14. Rowe CR, Zarins B. Recurrent transient subluxation of the shoulder. *J Bone Joint Surg Am.* 1981;63(6):863-872.
15. O'Brien SJ, Pagnani MJ, Fealy S, McGlynn SR, Wilson JB. The active compression test: a new and effective test for diagnosing labral tears and acromioclavicular joint abnormality. *Am J Sports Med.* 1998;26(5):610-613.
16. Tokish JT, Krishnan SG, Hawkins RJ. Clinical examination of the overhead athlete: the "differential-directed" approach. In: Krishnan SG, Hawkins RJ, Warren RF, eds. *The Shoulder and the Overhead Athlete.* Philadelphia, PA: Lippincott Williams & Wilkins; 2004:23-49.

II

*General
Imaging*

2

GENERAL IMAGING
OF THE SHOULDER

George Paul Hobbs, MD and William B. Morrison, MD

INTRODUCTION

Diagnostic imaging is an invaluable adjunct to clinical evaluation in the assessment of many upper extremity disorders. Knowledge of available imaging modalities is essential, allowing the most appropriate and cost-effective imaging test to be performed for a given clinical question. Ordering the correct modality with pertinent clinical details in the appropriate setting will yield maximum information facilitating optimum patient management.

Cohen SB. *Musculoskeletal Examination of the Shoulder:*
Making the Complex Simple (pp. 24-50).
© 2011 SLACK Incorporated

Radiography should be considered in all cases as a first step; yet, advanced imaging techniques such as ultrasound (US), computed tomography (CT), magnetic resonance imaging (MRI), and nuclear medicine are invaluable in the appropriate clinical setting. This chapter will first describe the modalities used in shoulder imaging and then focus on how imaging aids diagnosis with respect to specific clinical scenarios.

IMAGING MODALITIES

Radiography (x-ray)

Basic radiography has evolved from film-screen radiography to increasingly available digital and computed radiography techniques that are more amenable to manipulation, image transfer and storage, and viewing on workstations. Correctly performed views allow assessment of glenohumeral and acromioclavicular alignment as well as osseous structures at the shoulder. Radiographs are of particular value in the initial assessment of trauma where fracture or dislocation is suspected and in the assessment of arthritis, but they also prove useful in other cases such as tumor evaluation and rotator cuff calcific tendinosis.

Arthrography

Shoulder arthrography still has a role in the assessment of rotator cuff tears and adhesive capsulitis, although now it is rarely used in isolation since it is typically combined with MRI (direct MR arthrography) or CT (CT arthrography). MR arthrography is usually performed in cases where labral pathology is suspected and is of particular use in patients with instability.[1] For shoulder arthrography, a spinal needle (20 to 22 gauge) is used to gain access to the shoulder joint using an anterior or posterior approach using fluoroscopic guidance. Intra-articular position is confirmed with injection of a small amount of iodinated contrast, which is typically followed by 14 cc of dilute gadolinium when performing MR arthrography. Undiluted iodinated contrast is injected when performing CT arthrography.

Computed Tomography

The latest generation CT scanner uses multiple detector row arrays. Multidetector CT (MDCT) is a major improvement in CT technology, allowing simultaneous acquisition of multiple slices. This also enables acquisition of thinner slices than was previously possible, facilitating generation of exquisite multiplanar reformats and 3-dimensional (3-D) reconstructions. In the setting of metal hardware, this technology also results in remarkably decreased artifact. CT allows precise characterization of fractures at the shoulder, defining degree of comminution and displacement and intra-articular osseous bodies if present. CT arthrography is an alternative to MR arthrography, allowing assessment of rotator cuff tears (especially full thickness), labral tears, and cartilage loss at the glenohumeral joint.[2] CT may also be used for image-guided intervention.

Ultrasound

US is the medical imaging modality used to acquire and display the acoustic properties of tissues. A transducer array (transmitter and receiver of US pulses) sends sound waves into the patient and receives returning echoes with resulting data converted into an image. A high-frequency beam having a smaller wavelength provides superior spatial resolution and image detail. However, high-frequency energy is quickly absorbed by tissues; therefore, a lower frequency transducer may be needed for deeper structures. Thus, use of an appropriate transducer is of critical importance in performing shoulder imaging. Transducers up to 15 MHz are now available for high-resolution imaging of more superficial structures.

The acquisition of high-quality US images depends on operator experience, but in the right hands, US can be a powerful tool in the assessment of a wide range of shoulder pathology.[3-5] US is best used when a clinical question is well-formulated and the condition is dichotomous (eg, is there a full-thickness rotator cuff tear?). US is also excellent for assessment of injuries that are only observed during certain motions or in certain positions. Performing percutaneous interventions with US ensures accurate needle tip placement and helps direct the needle away from other regional soft tissue structures and neurovascular bundles.[6] Applications of US at the shoulder are depicted in Table 2-1.

Table 2-1

APPLICATIONS OF SHOULDER ULTRASOUND

- Rotator cuff integrity assessment (especially postarthroplasty)
- Evaluation and fenestration of rotator cuff calcific tendinosis and hypervascularity
- Evaluation of rotator cuff muscle atrophy
- Dynamic imaging of impingement
- Assessment of glenohumeral effusion
- Guided subacromial-subdeltoid bursal or joint injection

Magnetic Resonance Imaging

MRI is the workhorse of advanced musculoskeletal imaging throughout the body, and the shoulder is no exception. MRI relies on the magnetic properties of the hydrogen proton and does not involve ionizing radiation. It uses a strong magnetic field (most commonly 1.5 Tesla) to align these protons. Their energy level is altered by a transmitted radiofrequency (RF) pulse, after which they are allowed to relax back to their original state, releasing energy. This energy is used to formulate an image based on the different relaxation properties of tissues. Various "sequences" are prescribed to focus on the different relaxation properties of tissues in question. For instance, a "T1-weighted" sequence generally shows bright fat and dark fluid and is especially useful for depicting anatomy and differentiating hematopoietic (which contains fatty cells) marrow from tumor. On "T2-weighted" sequences, generally, fat is darker and fluid is bright, which is very useful for detecting pathology such as ligament and tendon tears. However, fat can be suppressed, so fluid is a better sign of which sequence has been used. There are other sequences that make fluid bright, so a more generic term, *fluid sensitive*, is used for these types of sequences. If gadolinium contrast is given intravenously or into the joint, T1-weighted sequences are used, and these images are generally fat suppressed to make the contrast more conspicuous. Gradient echo sequences

Table 2-2

MAGNETIC RESONANCE IMAGING CONTRAINDICATIONS*

Absolute

- Cardiac pacemaker/implantable cardiac defibrillators
- Ferromagnetic central nervous system (CNS) clips
- Orbital metallic foreign body
- Electronically, magnetically, and mechanically activated implants

Relative

- Prosthetic heart valves
- Foreign bodies in nonvital locations
- Infusion pumps/nerve stimulators
- Cochlear and stapedial implants

*Note: This is not intended as a complete list, but rather as a guide to MR contraindications. In individual cases, specific queries should be discussed directly with the radiology department before consideration for MR.

are commonly used because they are fast and can be used to acquire thinner slices. They can be T1 weighted or fluid sensitive but are prone to artifact, especially related to metal. Limitations of MRI include patient contraindications (Table 2-2), metallic susceptibility artifact from hardware, and patient intolerance secondary to claustrophobia. Open MR scanner configurations are available but are typically of lower field strength, thereby resulting in a lower quality image.

The principal applications of MR of the shoulder are shown in Table 2-3. Direct MR arthrography is best for evaluation of suspected labral tear. It involves direct injection of dilute gadolinium under fluoroscopic guidance as described above. It is also of value in assessing the rotator cuff. Indirect MR arthrography involves an intravenous injection of gadolinium that diffuses into the joint cavity, causing an "arthrographic" effect.

Table 2-3

SHOULDER MAGNETIC RESONANCE IMAGING APPLICATIONS

- Rotator cuff pathology
- Labral/capsular injury (best using direct MR arthrography)
- Rotator cuff interval lesions
- Muscle and other tendon tears
- Nerve pathology and impingement
- Osseous and soft tissue neoplasms
- Infection

Nuclear Medicine

In nuclear medicine exams, a radiopharmaceutical (radio-active isotope coupled to a pharmaceutical) is administered (usually intravenously), and subsequent gamma ray emission is detected by a gamma camera. An isotope whole body bone scan is used to detect areas of increased bone turnover, which may indicate pathology depending on the site and degree of activity. The main application in the shoulder is in the evaluation of neoplasia (particularly to assess for multiple foci around the body that would indicate metastatic disease) and osteomyelitis.[7] These exams are limited by inherent low resolution and low specificity.

The most significant advancement in nuclear medicine recently is positron emission tomography (PET) and combination PET CT scanners with important implications for oncology. [^{18}F] 2-deoxy-2-fluoro-D-glucose (FDG) is a metabolic tracer most widely used in clinical PET oncology. PET applications are evolving, but it is now approved for the diagnosis, staging, and restaging of many common malignancies and has shown efficacy for the detection of osseous metastasis from several malignancies, including lung carcinoma, breast carcinoma, and lymphoma.[8]

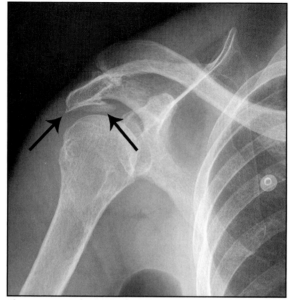

Figure 2-1. Subacromial spur. Frontal view of the right shoulder demonstrates a large osseous excrescence along the undersurface of the acromion (arrows).

CLINICAL INDICATIONS

Impingement

Radiographs are the first step in imaging evaluation of clinically suspected impingement. Findings that can be seen include subacromial spurs (Figure 2-1), greater tuberosity bone proliferation, high-riding humeral head (superior subluxation) (Figure 2-2), and glenohumeral osteoarthritis. In addition, an os acromiale (unfused ossification center at the anterior acromion, which is highly associated with cuff impingement) can be seen (Figure 2-3). Radiographs are also excellent at detecting calcific tendinosis and bursitis, which is very common at the shoulder (Figure 2-4). Other conditions that can simulate cuff pathology (ie, acromioclavicular joint osteoarthritis and clavicular osteolysis) are also seen (Figure 2-5).

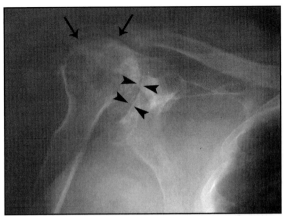

Figure 2-2. High-riding humeral head. Frontal view of the shoulder in a patient with rheumatoid arthritis demonstrates superior subluxation of the humeral head relative to the glenoid with abutment and remodeling of the acromial undersurface (arrows). The presence of marked joint space narrowing (arrowheads) is indicative of diffuse cartilage loss, typical for advanced rheumatoid arthritis.

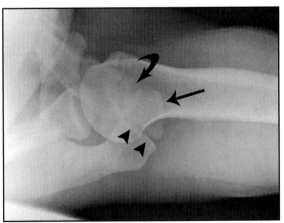

Figure 2-3. Os acromiale. Transaxillary view of the shoulder demonstrates an unfused ossification center (straight arrow) at the distal aspect of the acromion, bordered by the unfused acromial physis (arrowheads) and the acromioclavicular joint (curved arrow). Findings are typical for an os acromiale.

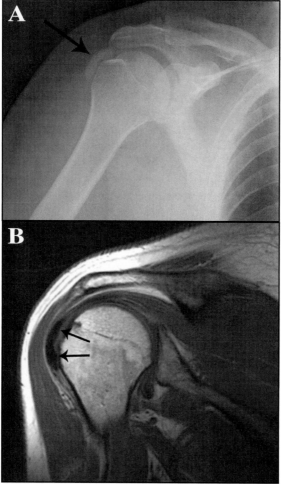

Figure 2-4. Calcific tendinopathy/bursitis. (A) Frontal x-ray of the shoulder demonstrates curvilinear calcium deposition (arrow) along the superolateral aspect of the humeral head in the expected region of the rotator cuff and overlying subacromial-subdeltoid bursa. (B) T1-weighted and (C) T2-weighted fat-suppressed coronal images of the same shoulder demonstrate low T1 and T2 signal calcium deposition (arrows) along the superficial aspect of the distal supraspinatus tendon, which extends into the overlying subacromial-subdeltoid bursa with associated edema/inflammation of the bursa (arrowheads) *(continued)*.

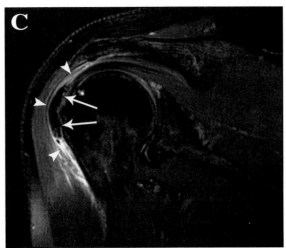

Figure 2-4 (continued). Calcific tendinopathy/bursitis. (B) T1-weighted and (C) T2-weighted fat-suppressed coronal images of the same shoulder demonstrate low T1 and T2 signal calcium deposition (arrows) along the superficial aspect of the distal supraspinatus tendon, which extends into the overlying subacromial-subdeltoid bursa with associated edema/inflammation of the bursa (arrowheads).

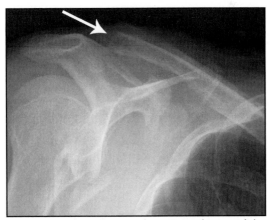

Figure 2-5. Clavicular osteolysis. Frontal x-ray of the right shoulder in a patient with hyperparathyroidism demonstrates widening of the acromioclavicular joint with bony resorption of the distal clavicle (arrow), consistent with clavicular osteolysis. The left clavicle (not shown) had a similar appearance.

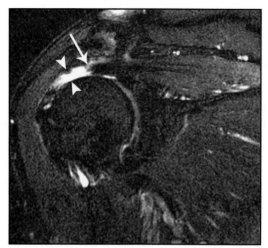

Figure 2-6. Rotator cuff tear by MRI. Coronal fat-suppressed T2-weighted image of the shoulder demonstrates a full-thickness tear of the distal supraspinatus tendon with medial retraction of the tendon fibers (arrow) and fluid in the expected location of the supraspinatus tendon (arrowheads).

The next step in imaging evaluation generally includes the use of MRI or US. MRI gives excellent detail and can identify rotator cuff tears and characterize the size and morphology of the tear as well as presence of muscle atrophy, spur morphology, and associated findings such as glenohumeral cartilage loss and superimposed labral tear. In this regard, MRI gives a global evaluation of the joint and can help differentiate different types of impingement (eg, "primary" impingement due to subacromial spur versus "secondary" impingement due to instability). It can help guide medical and surgical decision making in determining rehabilitation, subacromial decompression, cuff repair, or other options (Figure 2-6).

MR arthrography is usually not needed for detection of rotator cuff tears. Plain arthrography, however, remains an option if the clinical question is limited to whether there is a full-thickness tear; contrast injected into the joint will flow into the subacromial-subdeltoid bursa through the tear. However, it can be difficult to detect the size and location of the tear. Imaging with MDCT after arthrography facilitates this and can be a useful alternative if MRI is contraindicated (Figure 2-7).

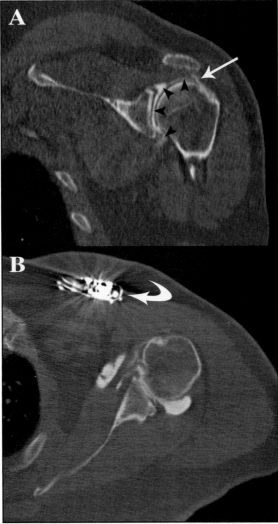

Figure 2-7. Rotator cuff tear by CT arthrogram. (A) Coronal oblique CT image of the left shoulder demonstrates high attenuation contrast material within the glenohumeral joint space (arrowheads) with contrast insinuating into a linear defect within the overlying supraspinatus tendon (straight arrow), consistent with a rotator cuff tear. (B) Axial CT image of the left shoulder in the same patient demonstrates a pacemaker (curved arrow) within the anterior left chest wall, which is a contraindication to MRI.

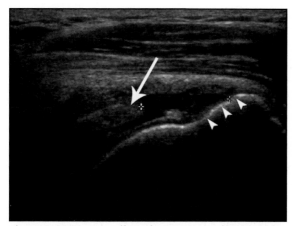

Figure 2-8. Rotator cuff tear by US. Coronal sonographic image of the superior shoulder (longitudinal to supraspinatus) demonstrates a full-thickness tear of the distal supraspinatus tendon with proximal retraction of the torn tendon (arrow) from the cortex of the greater tuberosity (arrowheads). The proximal and distal margins of the rotator cuff defect are delineated by crosses.

US is a very useful and cost-effective option for evaluating the rotator cuff and is highly sensitive at detection of rotator cuff tears and associated pathology, such as calcific tendinosis and muscle atrophy, similar to MRI. However, it does not give a global evaluation of the joint, and, in the United States, it is often reserved for situations where MRI is not feasible or optimal (Figure 2-8).

Instability/Suspected Labral Tear

Imaging of a patient with instability begins with radiographic evaluation. Radiographs can detect glenohumeral malalignment as well as numerous findings associated with instability such as Hill-Sachs defects, Bankart fractures, and greater tuberosity avulsion fracture related to dislocation, as well as glenoid deformity or osteoarthritis related to chronic instability (Figures 2-9 and 2-10).

However, radiographs alone rarely provide adequate information and cannot assess the status of the capsulolabral complex that is typically the source of the patient's symptoms.

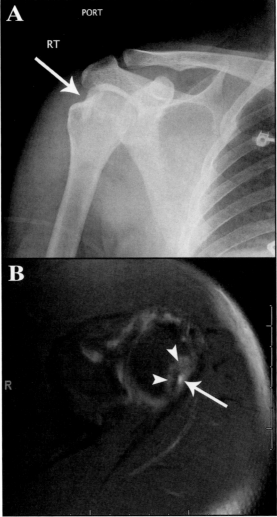

Figure 2-9. Hill-Sachs lesions. (A) Frontal x-ray of the right shoulder during internal rotation demonstrates a large defect in the posterolateral aspect of the humeral head (arrow), consistent with a chronic Hill-Sachs lesion secondary to remote anterior dislocation. (B) Axial T2-weighted fat-suppressed images of the shoulder at the superior aspect of the humeral head demonstrate a depressed fracture of the superior posterior aspect of the humeral head (arrow) with adjacent marrow edema (arrowheads), consistent with an acute Hill-Sachs lesion.

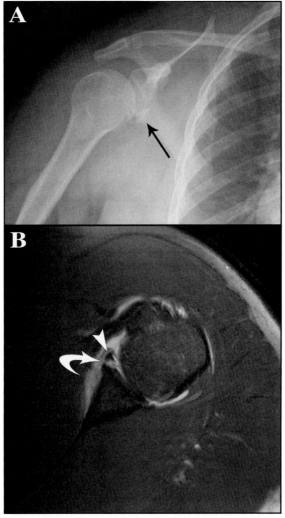

Figure 2-10. Osseous and soft tissue Bankart lesions. (A) Frontal x-ray of the shoulder demonstrates a fracture of the anteroinferior rim of the glenoid (straight arrow), consistent with an osseous Bankart lesion related to a recent anterior shoulder dislocation. (B) Axial T2-weighted image of the shoulder at the level of the inferior glenoid demonstrates a detached labral tear (curved arrow) with anterior displacement of the anterior-inferior labrum (arrowhead), consistent with a soft tissue Bankart lesion.

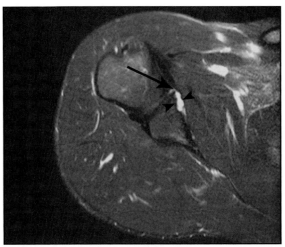

Figure 2-11. Labral tear with paralabral cyst. Axial T2-weighted image of the right shoulder demonstrates a focal tear of the anteroinferior labrum (arrow) with joint fluid extravasating through the labral defect and forming a paralabral cyst (arrowheads).

The best test for this is MRI, especially MR arthrography. Injection of gadolinium contrast into the joint followed by MRI yields excellent depiction of anatomy, and distention of the joint with fluid is very useful in demonstrating abnormal communication through labral and capsular attachments. The same is true for evaluation of superior labral tear (SLAP tear). Paralabral cysts, which signify underlying labral tear and occasionally impinge upon adjacent nerve branches, are also easily detected (Figure 2-11).

If arthrography is not feasible, noncontrast MRI remains highly sensitive and specific. If MRI is contraindicated, CT arthrography using MDCT is also very accurate. US is limited for evaluation of capsulolabral pathology.

If multidirectional instability (MDI) is suspected, MR arthrography remains the best test, although generally all that is seen is capsular laxity with large recesses and prominent fluid in the rotator cuff interval. Nevertheless, MR arthrography is useful to exclude an anatomic lesion that may benefit from surgical repair rather than rehabilitation.

Frozen Shoulder

In patients with limited or painful range of motion, adhesive capsulitis is a diagnostic consideration. This can occur following injury or immobilization but is often seen in association with other pathology including rotator cuff tears. A common finding on MRI and MR arthrography is synovitis, edema, and a mass effect at the rotator cuff interval between the supraspinatus and subscapularis (Figure 2-12). Alternatively, US with color Doppler can detect hyperemic tissue at the rotator cuff interval.

Acute Trauma

Muscle Injury

For evaluation of muscle injury at the shoulder, MRI is generally the best option, yielding global evaluation of the musculature as well as the underlying joint. It should be noted that the clinician should always include the suspected diagnosis on the MRI prescription because imaging protocol depends on the suspected injury; for rotator cuff muscle injury, a standard shoulder MRI is optimal, whereas evaluation of pectoralis major tear requires a larger field-of-view including the medial soft tissues and insertion site on the humeral shaft (Figure 2-13). A biceps tear presenting with a lump on the upper arm can be due to a distal tear at the elbow or a proximal tear of the long head; the incorrect study to request is an "upper arm"—the relevant findings would be at the elbow or shoulder.

Fracture/Dislocation

For acute osseous injury, radiographs are essential. CT may be needed to evaluate complex fractures or those involving the articular surfaces in order to plan surgery. Reformatted 2-D and 3-D images can also be acquired. For suspected acromio-clavicular joint injury, it can be useful to include stress views using counterweights. MRI can be used to assess status of the coracoclavicular ligaments if warranted (Figure 2-14). As listed previously under "Instability," if glenohumeral joint dislocation is suspected, radiographs are essential, but MRI is often required to assess capsulolabral lesions.

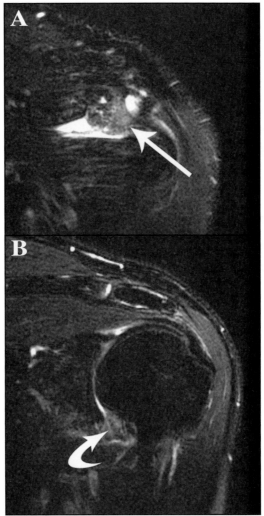

Figure 2-12. Adhesive capsulitis. (A, B) Coronal T2-weighted fat-suppressed images and (C) sagittal T1-weighted image of the shoulder in a patient with decreased range of motion demonstrate isointense synovial proliferation within the rotator cuff interval (straight arrows) and axillary recess (curved arrow), indicative of adhesive capsulitis *(continued)*.

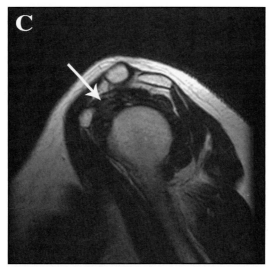

Figure 2-12 (continued). Adhesive capsulitis. (A, B) Coronal T2-weighted fat-suppressed images and (C) sagittal T1-weighted image of the shoulder in a patient with decreased range of motion demonstrate isointense synovial proliferation within the rotator cuff interval (straight arrows) and axillary recess (curved arrow), indicative of adhesive capsulitis.

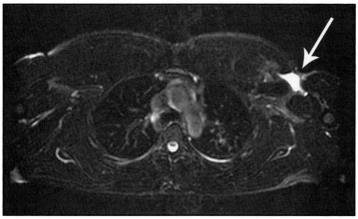

Figure 2-13. Pectoralis major tear. Axial T2-weighted image of the upper chest demonstrates a hematoma at the musculotendinous junction of the left pectoralis major (arrow), consistent with a tear.

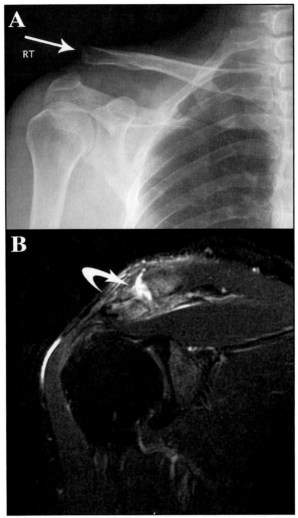

Figure 2-14. Acromioclavicular separation. (A) Frontal view of the right shoulder demonstrates widening of the acromioclavicular joint with superior displacement of the distal clavicle (straight arrow) relative to the acromion. (B) A T2-weighted image of the shoulder in a different patient demonstrates widening of the acromioclavicular joint with an associated joint effusion (curved arrow) and superior displacement of the distal clavicle relative to the acromion.

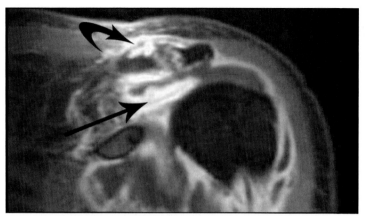

Figure 2-15. Septic arthritis. Postcontrast T1-weighted fat-suppressed image of the right shoulder demonstrates diffuse synovial thickening and enhancement about the glenohumeral joint (straight arrow), which extends through a rotator cuff defect into the acromioclavicular joint (curved arrow). A glenohumeral joint aspirate was cultured and grew *Staphylococcus aureus*.

Infection

In the setting of a clinically suspected infection, radiographs are the best first-imaging exam. However, it should be recognized that osseous and articular findings associated with osteomyelitis and septic arthritis (joint effusion, erosions, periosteal reaction, and frank bone destruction) will not be apparent on radiographs until the infection is advanced. Therefore, cross-sectional imaging (especially MRI without and with contrast) is indicated in order to identify the location of the process as well as fluid loculations or bone marrow abnormalities that are best to target for sampling (Figure 2-15). Bone scanning is of limited utility in this regard due to lack of resolution. US can be very useful for detection (and percutaneous aspiration) of fluid collections but cannot evaluate extent of involvement in the bone.

Arthritis

Degenerative

Radiographs are generally adequate for evaluation of osteoarthritis. Classic hallmarks include joint space narrowing,

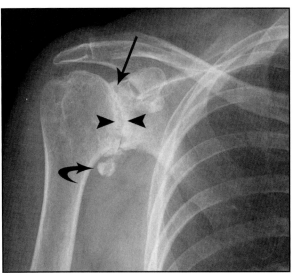

Figure 2-16. Osteoarthritis. Frontal x-ray of the right shoulder demonstrates marked glenohumeral joint space narrowing with subchondral sclerosis (arrowheads) and subchondral cysts (straight arrow). Intra-articular ossific bodies are present within the axillary recess (curved arrow).

osteophytes (spurs), and subchondral cysts or sclerosis (Figure 2-16). Cross-sectional imaging including MRI, CT, or US is useful depending on the clinical setting. For example, prior to total shoulder arthroplasty, US or MRI can be useful to evaluate the status of the rotator cuff, and CT can be useful to assess glenoid version and bone stock.

Inflammatory/Other Etiology

For arthropathies that are not related to mechanical origin, radiographs remain useful as the primary modality; erosions, calcifications, and other findings can be used in association with overall clinical assessment to arrive at a differential diagnosis. Advanced imaging can be used for problem solving. For example, CT or MRI can detect calcifications related to synovial osteochondromatosis (Figure 2-17). MRI can detect hemorrhage in conditions such as hemophilia and pigmented villonodular synovitis (Figure 2-18).

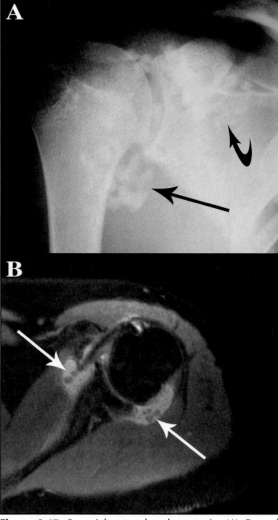

Figure 2-17. Synovial osteochondromatosis. (A) Frontal x-ray of the shoulder demonstrates multiple ossific bodies within the axillary recess (straight arrow) and subcoracoid recess (curved arrow). (B) Axial T1-weighted image of the shoulder after the intra-articular administration of contrast demonstrates intra-articular bodies within the anterior and posterior joint recesses (arrows).

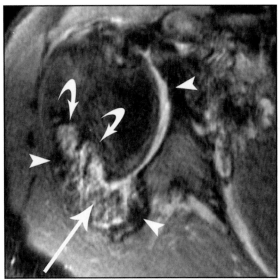

Figure 2-18. Hemophilic arthropathy. T2-weighted axial image demonstrates blood products within the joint space (straight arrow), low signal synovium secondary to hemosiderin deposition (arrowheads), and marked erosive changes of articular surface (curved arrows).

Tumor

Bone Tumor

Evaluation of a bone tumor always begins with radiographs. Radiographs can evaluate aggressive versus nonaggressive features. Nonaggressive features are associated with slow growth and benign lesions, the most reliable of which is a thin, sclerotic margin. A well-defined lesion without a thin, sclerotic margin cannot be assumed to be benign and includes metastases and myelomatous lesions. Aggressive characteristics include ill-defined margins, periosteal reaction, cortical breakthrough, and soft tissue mass effect. However, it should be recognized that aggressive lesions are not necessarily malignant, and some benign lesions such as infection can have these findings. Radiographs and CT can be used for characterization of these findings as well as matrix mineralization, and an assessment of aggressiveness as well as a limited differential diagnosis can be generated. MRI with and without contrast is very useful for problem solving, such as for evaluation of solid versus cystic nature and for determining areas best for biopsy. Whole body bone scan is generally useful only for detection of other lesions indicating mutifocality in cases of metastatic disease, disseminated infection, metabolic conditions, and other situations. Amount of uptake in a lesion is not a reliable indicator of benign or malignant etiology.

Soft Tissue Mass

For evaluation of a soft tissue mass, MRI is the optimal test. Size, location, and association with fascial planes and neurovascular structures can be assessed, and a global evaluation of the surrounding soft tissues and adjacent joint can be performed. Often, a specific diagnosis can be rendered (eg, lipoma, hemangioma) but even if findings are nonspecific and biopsy is required, MRI provides a roadmap. Intravenous contrast can be useful for detection of cystic or necrotic areas. Radiographs can be useful as an adjunct to detect patterns of calcification; however, calcification rarely provides a specific diagnosis and does not preclude the need for MRI. CT or US can be performed if MRI is contraindicated, but each has limitations. CT has limited specificity in terms of tissue characteristics, and US cannot provide global evaluation of the surrounding tissues. However, either can be used for imaging-guided biopsy if warranted.

An important caveat for all tumors, especially primary tumors, is that if biopsy is considered, consultation with an orthopedic tumor surgeon should be performed first to plan the approach in order to avoid contamination of overlying soft tissues outside of the definitive surgical field.

Imaging Considerations in the Postoperative Shoulder

Considering the multitude of procedures that are performed on the shoulder and the potential effect on subsequent image quality, it is important for the referring clinician to determine what type of surgery has been performed prior to requesting further imaging. MRI is most adversely affected by

Table 2-4

ANTICIPATED METAL ARTIFACT ON MAGNETIC RESONANCE IMAGING AND RECOMMENDATIONS

Surgery/Implant	Artifact on Magnetic Resonance Imaging	Recommendation
Bioabsorbable implant (radiolucent)	None to mild	No change in protocol needed
Osteotomy (eg, distal clavicular resection)	None to mild	No change in protocol needed
Metal anchors (cuff/labral repair)	Mild to moderate	May require alteration of protocol
Rod/plate	Moderate; less if titanium	Alter protocol or use alternate modality
Screws	Moderate to severe	Alter protocol or use alternate modality
Burring (eg, acromioplasty-microscopic metal)	Moderate to severe	Alter protocol or use alternate modality
Shrapnel/bullet fragments	Severe	Alter protocol or use alternate modality
Prosthesis	Severe	Use alternate modality

surgery, and exam quality can for the most part be predicted by knowledge of type of surgery. Preliminary radiographs can be used to determine potential for metal artifact if surgical history is not clear. Table 2-4 can be used to determine when an alternate modality should be considered. If MRI is contraindicated, US or CT arthrography may be helpful depending on the clinical question.

REFERENCES

1. Magee T, Williams D, Mani N. Shoulder MR arthrography: which patient group benefits most? *Am J Roentgenol.* 2004;183(4):969-974.
2. Lecouvet FE, Simoni P, Koutaissoff S, et al. Multidetector spiral CT arthrography of the shoulder: clinical applications and limits, with MR arthrography and arthroscopic correlations. *Eur J Radiol.* 2008;68(1):120-136.
3. Kolla S, Motamedi K. Ultrasound evaluation of the shoulder. *Semin Musculoskelet Radiol.* 2007;11(2):117-125.
4. Moosikasuwan JB, Miller TT, Burke BJ. Rotator cuff tears: clinical, radiographic, and US findings. *Radiographics.* 2005;25(6):1591-1607.
5. Papatheodorou A, Ellinas P, Takis F, et al. US of the shoulder: rotator cuff and non-rotator cuff disorders. *Radiographics.* 2006;26(1):e23.
6. Joines MM, Motamedi K, Seeger LL, DiFiori JP. Musculoskeletal interventional ultrasound. *Semin Musculoskelet Radiol.* 2007;11(2):192-198.
7. Schauwecker DS. The scintigraphic diagnosis of osteomyelitis. *Am J Roentgenol.* 1992;158(1):9-18.
8. Even-Sapir E. PET/CT in malignant bone disease. *Semin Musculoskelet Radiol.* 2007;11(4):312-321.

III

Common
Conditions of the
Shoulder

3

GLENOHUMERAL INSTABILITY

Steven B. Cohen, MD

INTRODUCTION

Shoulder instability is defined as either glenohumeral joint formal dislocation (requiring a reduction) or subluxation (transient shifting of the humeral head without frank dislocation). The 2 types of instability most commonly described are traumatic (single episode) or atraumatic (without a known cause, or microtraumatic). The direction of dislocation is determined by the mechanism of injury and describes the direction of the humeral head in relation to the glenoid. The most common direction of shoulder instability is anterior, which occurs while the arm is in the abducted and external

Cohen SB. *Musculoskeletal Examination of the Shoulder:*
Making the Complex Simple (pp. 52-73).
© 2011 SLACK Incorporated

rotation position and the hand or arm is pushed posteriorly or in extension. Posterior instability, which accounts for 2% to 10% of all shoulder instability,[1-3] is produced when a posterior-directed force is applied to the adducted, forward-flexed, and internally rotated arm.

As mentioned previously, the most common form of traumatic instability is anterior, while the most common atraumatic form is multidirectional instability (MDI). The main difference pathologically between unidirectional instability and MDI is global laxity of the capsule inferiorly, posteriorly, and anteriorly compared to one direction (eg, in anterior instability).

The age at the time of initial shoulder instability is possibly the biggest factor in the development of recurrent instability, particularly after a traumatic event.[4] Patients younger than 20 years old have a 40% to 90% chance of sustaining another dislocation or having additional subluxation events.[5-8] In addition, other risk factors that can predispose to recurrent instability include a tear of the anteroinferior labral complex (known as the Bankart lesion) and a humeral head indentation fracture (Hill-Sachs lesion).[9]

HISTORY

The recognition of a traumatic shoulder dislocation may be readily apparent; however, diagnosing subtle shoulder instability can be challenging. A thorough history and physical examination are critical in determining the correct diagnosis (Table 3-1). As mentioned above, determining the age of the patient's first instability event is paramount for the treatment. It is important to determine whether the instability was traumatic or atraumatic. In addition, the direction of instability can often be determined by questioning the patient as to his or her arm position during the instability event. It is also critical to know if the patient required a formal reduction of the dislocation under sedation or if a true subluxation event occurred. A formal dislocation requiring reduction typically requires more force and may potentially cause a greater acute injury (labral tear/Hill-Sachs lesion), whereas a lower energy subluxation event may be more indicative of capsular/ligamentous stretching.

Table 3-1

HELPFUL HINTS

Type of Instability

DIRECTION	DESCRIPTION/MECH-ANISM OF INJURY	TYPICAL PATIENT COMPLAINT
Anterior	Shoulder in abduction and external rotation with a posterior-directed force to the arm	Pain and apprehension with reaching away from body Inability to do overhead sports (throw, swim, volleyball)
Posterior	Shoulder in adduction, forward flexion, and internal rotation (provocative position) with a posterior-directed force to the arm	Pain and instability in provocative position Inability to do bench press or push-ups
Multidirectional	No traumatic injury, microtrauma (typically swimming), or multiple injuries with various injuries	Constant sensation of instability Dislocations while asleep or with minimal force Weakness

Imaging

IMAGE	PERTINENT IMAGE VIEWS	FINDINGS
Plain radiograph (x-rays)	Axillary view, Stryker notch view	Evaluated direction of dislocation, bony injuries (Hill-Sachs lesion, bony Bankart injury, glenoid fracture)
Magnetic resonance imaging (MRI—arthrogram)	Axial, coronal	Evaluate for labral and capsular injuries
Computed tomography (CT scan)	Axial, coronal	Evaluate for bony injuries (especially large Hill-Sachs lesions)

After the initial or recurrent event and after the acute pain subsides, it is important for the patient to regain his or her range of motion (ROM) and strength. Once the initial pain from the episode has resolved, the patient may complain of shoulder shifting (instability), difficulty with overhead activity, inability to do sports, instability while sleeping, trouble with activities of daily living, episodes of "dead arm" sensation, and/or failed prior attempts at physical therapy.

EXAMINATION

The physical examination of the unstable shoulder is similar for other pathologic conditions of the shoulder, including inspection, palpation, ROM testing, strength testing, and impingement evaluation (Table 3-2). However, additional tests must be performed to assess the unstable shoulder. It is important to inspect the shoulder girdle for muscle atrophy consistent with a nerve injury or compression. An injury to the axillary or musculocutaneous nerve may occur with a traumatic dislocation. This may present as weakness or poor deltoid function and decreased sensation over the lateral aspect of the shoulder in an axillary nerve palsy and elbow flexion weakness and decreased sensation over the lateral forearm in the case of a musculocutaneous nerve injury. In addition, in the case of chronic or recurrent instability, inspection of the scapula is crucial to evaluate for scapular dyskinesia as evident by scapular winging, scapular protraction, and scapular depression or a scapular malposition, inferior medial scapular winging, coracoid tenderness, and scapular dyskinesis (SICK) scapula.

The shoulder is then palpated for tenderness specifically over the acromioclavicular (AC) joint and biceps tendon. Tenderness over the AC joint and pain with horizontal adduction of the shoulder may be indicative of AC joint arthrosis. Bicipital groove and biceps tendon tenderness in association with pain with Speed's and Yergason's tests are consistent with biceps tendonitis or other biceps tendon pathology. It is also important to palpate the cervical spine to rule out any cervical pathology as well as the medial border of the scapula to rule out scapulothoracic bursitis.

Table 3-2

METHODS FOR EXAMINING THE UNSTABLE SHOULDER

Examination	Technique	Illustration	Grading	Significance
Range of motion	Examine active and passive motion	1. External rotation  2. Forward flexion 		Loss of active motion may be related to pain Loss of passive motion may be from capsular contracture

(continued)

Table 3-2 (continued)

METHODS FOR EXAMINING THE UNSTABLE SHOULDER

Examination	Technique	Illustration	Grading	Significance
Strength testing	Manual strength testing: Internal and external rotation, abduction, scaption		Grade 5 = full strength Grade 4 = < full strength Grade 3 = movement with gravity Grade 2 = movement without gravity Grade 1 = visible contraction	Loss of strength can indicate rotator cuff pathology or nerve injury
Ligamentous laxity	Tests: Thumb to forearm; passive MCP extension of small finger >90 degrees, elbow hyperextension; knee hyperextension, sulcus sign, palms flat to floor	Thumb to forearm	Beighton Ligamentous Laxity Scale 0 to 3 = tight 4 to 6 = hypermobile 7 to 9 = excessively hypermobile	Positive finding may predispose to shoulder instability

(continued)

Table 3-2 (continued)

METHODS FOR EXAMINING THE UNSTABLE SHOULDER

Examination	Technique	Illustration	Grading	Significance
Impingement signs	Neer: Passive flexion >90 degrees with arm internal rotation and scapula stabilized Hawkins: Internal rotation with arm flexed to 90 degrees	Neer Hawkins 	Positive pain = impingement	Impingement may be a sign of rotator cuff tendonitis or internal impingement in throwers
Apprehension sign	90 degrees of external rotation and abduction		Positive if patient senses anterior instability	Indication of anterior instability

(continued)

Table 3-2 (continued)

METHODS FOR EXAMINING THE UNSTABLE SHOULDER

Examination	Technique	Illustration	Grading	Significance
Relocation test	Same as apprehension with posterior-directed force on humerus		Positive if sensation of more stable shoulder	Indication of anterior instability
Axial load and shift test	Patient supine with arm abducted and axial force applied to humerus: Shift anterior and posterior		1+ = translation to glenoid rim 2+ = translation over glenoid rim 3+ = frank dislocation	Positive anterior displacement = anterior instability Positive posterior displacement = posterior instability
Sulcus sign	Downward-directed force of humerus with arm at the side in neutral and 90 degrees of external rotation		1+ = 0 to 1 cm 2+ = 1 to 2 cm 3+ = >2 cm	A 3+ exam that remains 2+ in external rotation is pathognomonic for multidirectional instability

(continued)

Table 3-2 (continued)

METHODS FOR EXAMINING THE UNSTABLE SHOULDER

Examination	Technique	Illustration	Grading	Significance
O'Brien test	1. Shoulder to 90 degrees flexion, 30 degrees of adduction, and thumb pointing downward with resisted forward flexion 2. Rotate to full supination and resist forward flexion	1.  2.	Positive = pain or click	1. Pain or click = SLAP tear 2. Anterosuperior pain = acromioclavicular joint pathology
Jerk test	Shoulder abducted and flexed to 90 degrees and brought to adduction	A B Kim SH, Park JC, Park JC, Oh I. *Am J Sports Med* (Vol. 32, Iss. 8) pp. 1849-1855, copyright © 2004 by SAGE Publications. Reprinted with permission of SAGE Publications.	Positive = posterior pain and clunk	Indicator of posteroinferior instability

(continued)

Table 3-2 (continued)

METHODS FOR EXAMINING THE UNSTABLE SHOULDER

Examination	Technique	Illustration	Grading	Significance
Kim test	Shoulder flexed to 135 degrees and posterior and inferior force is applied to proximal arm	 Kim SH, Park JS, Jeong WK, Shin SK. *Am J Sports Med* (Vol. 33, Iss. 8) pp. 1188-1192, copyright © 2005 by SAGE Publications. Reprinted with permission of SAGE Publications.	Positive = posterior clunk/pain	Indicator of posteroinferior instability
Circumduction test	Arm started in adduction and brought into abduction and external rotation	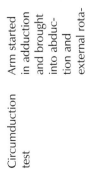	Postitive = posterior clunk	Indicator of posterior instability

(continued)

Table 3-2 (continued)

METHODS FOR EXAMINING THE UNSTABLE SHOULDER

Examination	Technique	Illustration	Grading	Significance
Speed's test	Shoulder flexion with elbow in full extension and forearm in full supination		Positive = pain	Indicates biceps tendonitis or SLAP tear

Instability Magnetic Resonance Imaging Findings

STRUCTURE	MAGNETIC RESONANCE IMAGING FINDING
Anterior labrum	± Bankart tear
Posterior labrum	± Labral tear
Superior labrum	± SLAP tear
Inferior capsule	± Capsular laxity, enlarged axillary pouch
Humeral head	± Hill-Sach's/reverse Hill-Sach's lesion
Glenoid	± Bony Bankart or glenoid erosion

(continued)

Table 3-2 (continued)

METHODS FOR EXAMINING THE UNSTABLE SHOULDER

Pertinent Anatomy in Shoulder Instability

STRUCTURE	FUNCTION	IMPORTANCE
Inferior glenohumeral ligament	Anterior band: Resists anterior translation with the arm in 90 degrees of abduction and external rotation	Injury or stretch allows increased anterior or posterior translation of the humeral head
	Posterior band: Resists posterior translation with the arm in 90 degrees of forward flexion and internal rotation	
Middle glenohumeral ligament	Resists anterior translation with the arm abducted 45 degrees	Injury or stretch allows increased anterior translation of the humeral head
Superior glenohumeral ligament	Resists inferior translation with arm at side	Injury or stretch allows increased inferior translation of the humeral head
Labrum	Site of attachment anteriorly for glenohumeral ligaments	Tear or injury allows increased humeral head translation (usually associated with capsular stretch)
	Acts as chock block to deepen glenoid cavity	

(continued)

Table 3-2 (continued)

METHODS FOR EXAMINING THE UNSTABLE SHOULDER

Pertinent Anatomy in Shoulder Instability

STRUCTURE	FUNCTION	IMPORTANCE
Rotator interval	Capsule at the interval between subscapularis and supraspinatus that resists inferior translation. Includes the superior glenohumeral ligament, coracohumeral ligament, and joint capsule.	Injury or stretch allows increased inferior translation with the arm at the side in neutral and external rotation. May also allow increased posterior translation
Posterior capsule	Thin tissue with no ligamentous attachments. Resists posterior translation of the humeral head.	Injury or stretch allows increased posterior translation
Coracohumeral ligament	Ligamentous structure from the base of the coracoid to the greater tuberosity. Resists inferior translation.	Injury or stretch allows increased inferior translation with the arm at the side in neutral and external rotation

SLAP = superior labral anterior posterior

In chronic instability or typically shortly after an acute instability episode, ROM of the shoulder is returned to symmetric to the contralateral side. It is important to confirm that the active motion is similar to the passive motion. Any pain associated with particular motions or limitation should be noted. In addition to glenohumeral motion, scapulothoracic motion should be assessed for winging and scapular dyskinesia.

Following ROM testing, strength testing is performed to assess the rotator cuff and deltoid integrity. Testing of the rotator cuff includes evaluation of internal and external rotation strength at the side and elevation strength at 90 degrees in the plane of the scapula (scaption). Weakness in any of these may be associated with rotator cuff pathology or nerve compression. In addition, if there is an axillary nerve injury from an acute dislocation, there may be weakness of the anterior, middle, or posterior deltoid as indicated by weakness of shoulder forward elevation, abduction, and extension, respectively, with the elbow flexed at the side. Scapular muscle strength should also be tested by testing shoulder shrug, protraction, retraction, and wall push-ups.

Other tests for the unstable shoulder should include impingement tests of Neer and Hawkins as well as biceps tendon testing including Speed's and Yergason's tests. Generalized ligamentous laxity is usually measured using the Beighton scale.[10] This is performed by assessing for hyperextension of the knee, elbow, and metacarpophalangeal joint (MCP), as well as thumb-forearm testing and palms-to-floor hamstring stretching. Patients with positive ligamentous laxity testing may be more prone to the multidirectional type of shoulder instability.

Intuitively, stability testing is the most important portion of the exam for patients with shoulder instability. A "load and shift" maneuver, as described by Murrell and Warren, is performed with the patient supine, testing for anterior and posterior translation.[11] The arm is held in 90 degrees of abduction and neutral rotation while a posterior force is applied in an attempt to translate the humeral head over the posterior glenoid. Translation is graded "0" if the humeral head does not translate to the glenoid rim, "1+" if the humeral head translates to the glenoid rim and is increased compared to the uninvolved shoulder, "2+" if the humeral head translates over the glenoid rim but spontaneously reduces, and "3+" if the humeral head translates over the glenoid rim and does not spontaneously reduce. The sulcus sign is tested in neutral and external rotation for inferior translation in the seated position. A sulcus sign graded as 3+ that remains 2+ in external rotation is pathognomonic for multidirectional instability.[12] It is critical to test both shoulders, as symmetric laxity may be observed that may not be pathologic for the specific patient.

Other tests that are important in anterior instability include the apprehension test and the Jobe relocation test. These may also be positively associated with anterior and/or superior labral tears. Patients with symptomatic posterior instability may have positive findings with the jerk, circumduction, and Kim tests.[13,14] The O'Brien and Mayo Shear or grind tests can be positive with superior labral, biceps tendon, or AC joint pathology.

PATHOANATOMY

Advances into the pathology associated with shoulder instability have come as a result of improved imaging, arthroscopy, and anatomical and biomechanical studies. Injuries to the unstable shoulder can be classified as soft tissue or bony. Initial traumatic dislocations are associated most commonly with labral tears. The location of the tears depends on the direction of the dislocation. With more recurrent instability episodes, stretching of the capsuloligamentous structures occur. The anterior band of the inferior glenohumeral ligament (IGHL) is the primary restraint to anterior translation with the shoulder in abduction and external rotation, whereas the posterior band of the IGHL restricts posterior translation with the arm in forward flexion and internal rotation. The superior glenohumeral ligament and rotator interval restrict inferior translation of the shoulder with the arm at the side. With either acute or chronic instability, bony injury may occur. These injuries include Hill-Sachs or reverse Hill-Sachs lesions for anterior and posterior instability, respectively, as well as fractures to the glenoid rim (bony Bankart lesions). Some anatomical variants may also predispose to shoulder instability including glenoid hypoplasia and excessive humeral head retroversion.

IMAGING

Routine x-rays are necessary for the evaluation of any patient with a history of instability. The typical series includes anteroposterior (AP) view, axillary view, outlet view, and a Stryker notch view. However, it is essential to obtain x-rays in multiple planes (ie, an AP and axillary view) to confirm the reduction of an acutely dislocated shoulder. Radiographs are necessary to evaluate for Hill-Sachs lesions or glenoid pathology. The gold standard for assessment of the unstable shoulder is magnetic resonance imaging (MRI) with or without gadolinium arthrogram enhancement. This study allows for evaluation of the capsulolabral complex, which is most commonly injured with shoulder instability (Figure 3-1). In addition, MRI is useful for assessment of capsular laxity (Figure 3-2) and biceps tendon and rotator cuff lesions. Computed tomography (CT) may be useful for the evaluation of bony pathology such as larger Hill-Sachs lesions and glenoid fractures or hypoplasia.

TREATMENT

Conservative

The initial treatment of nearly all types of shoulder instability is nonoperative rehabilitation. The exceptions to this include locked or chronic dislocations that are not reduced and, in some cases, high-level contact athletes who are at high risk of developing recurrent instability. After an acute dislocation, the standard treatment is a sling for shoulder comfort in the initial painful period, which typically lasts several weeks. Once the acute pain has subsided, shoulder ROM is begun. Once full active and passive ROM is achieved, a strengthening program is initiated. As mentioned previously, younger patients are at higher risk for recurrent instability if the first dislocation occurred before the age of 20. Some athletes use a protective harness during athletic participation to protect themselves from reinjury. If instability persists despite vigorous attempts at nonoperative treatment, then surgical management is considered.

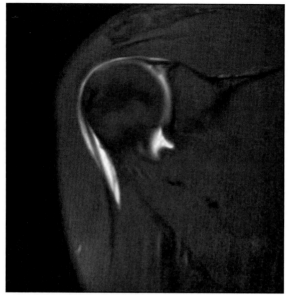

Figure 3-1. MRI coronal image of a SLAP tear.

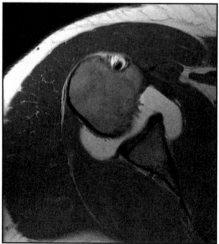

Figure 3-2. MRI axial image of capsular laxity seen in multidirectional instability.

Surgical

Anterior Instability

The surgical management of unidirectional anterior instability predominantly involves arthroscopic repair of the anterior-inferior labral tear or Bankart lesion (Figure 3-3). This may or may not include an anterior capsular shift to reduce the redundancy of the anterior capsule and IGHL that occurs from recurrent anterior dislocations. In select cases of some contact athletes, as well as those patients who have failed an arthroscopic repair, an open anterior capsular shift is performed. This had been considered the "gold standard" by which arthroscopy has been compared because most studies have a success rate of more than 90% after open labral repair and capsular shift.[15-17] However, an open surgery requires either detachment or a split of the subscapularis tendon, which can lead to some internal rotation weakness or dysfunction. With improved arthroscopic techniques, success rates have been close to open results,[18,19] thus, an arthroscopic repair is performed in the large majority of cases. Regardless of the repair method, most patients are unable to return to normal athletic participation for 4 to 6 months after shoulder surgery for instability.

Posterior Instability

Similar to anterior instability (except with anterior pathology), posterior instability commonly results in tearing of the posterior labrum and stretching of the thin posterior capsule and the posterior band of the IGHL (Figure 3-4). Recurrent symptoms present with instability in the provocative position of shoulder flexion, adduction, and internal rotation. If symptoms persist despite attempts at rehabilitation, then an arthroscopic labral repair and capsular shift can be performed similar to that of anterior instability.[1,11,19-21] This typically results in the reduction of symptoms and the ability to return to normal functional activities, including sports, after 4 to 6 months of rehabilitation.

Multidirectional Instability

The surgical treatment of MDI is approached differently than unidirectional anterior or posterior instability. MDI is

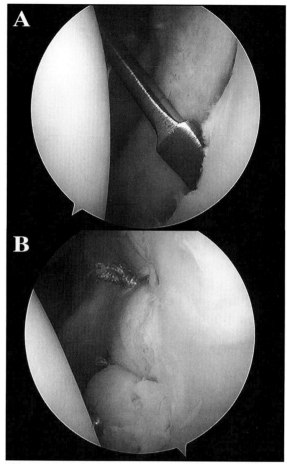

Figure 3-3. Arthroscopic image (A) before and (B) after anterior labral repair.

generally not the result of a traumatic injury and an associated labral tear. As a result, treatment requires a capsular tightening procedure (Figure 3-5), the direction of which is dependent on the symptoms and in particular the examination under anesthesia (EUA). If the EUA reveals significant instability in both anterior and posterior directions, then an arthroscopic capsular shift is performed in both directions.[22,23]

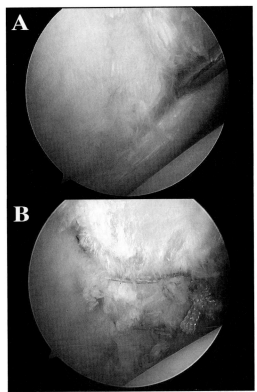

Figure 3-4. Arthroscopic image (A) before and (B) after posterior labral repair.

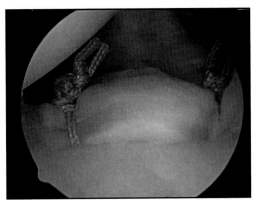

Figure 3-5. Arthroscopic image after a capsular shift for multidirectional instability.

If there is a positive sulcus sign that does not decrease when the arm is externally rotated at the side, then a rotator interval tightening is performed in conjunction with the capsular shift at the time of surgery to reduce the inferior instability.[24,25] The postoperative rehabilitation consists of treatment in a sling for 4 to 6 weeks with return to full unrestricted activity by 4 to 6 months.

Conclusion

The diagnosis of shoulder instability is based on a good history and a thorough physical examination. The proper treatment is based on the correct diagnosis, which is guided by all of the examination tools as described previously.

References

1. Antoniou J, Duckworth DT, Harryman DT II. Capsulolabral augmentation for the management of posteroinferior instability of the shoulder. *J Bone Joint Surg Am.* 2000;82(9):1220-1230.
2. Boyd HB, Sisk TD. Recurrent posterior dislocation of the shoulder. *J Bone Joint Surg Am.* 1972;54A:779.
3. McLaughlin HL. Posterior dislocation of the shoulder. *J Bone Joint Surg Am.* 1952;34A:584.
4. Hovelius L, Olofsson A, Sandström B, et al. Nonoperative treatment of primary anterior shoulder dislocation in patients forty years of age and younger: a prospective twenty-five-year follow-up. *J Bone Joint Surg Am.* 2008;90(5):945-952.
5. Gartsman GM, Roddey TS, Hammerman SM. Arthroscopic treatment of anterior-inferior glenohumeral instability: two to five-year follow-up. *J Bone Joint Surg.* 2000;82A:991-1003.
6. Hovelius L, Augustini BG, Fredin H, Johansson O, Norlin R, Thorling J. Primary anterior dislocation of the shoulder in young patients: a ten-year prospective study. *J Bone Joint Surg Am.* 1996;78(11):1677-1684.
7. Itoi E, Hatakeyama Y, Sato T, et al. Immobilization in external rotation after shoulder dislocation reduces the risk of recurrence: a randomized controlled trial. *J Bone Joint Surg Am.* 2007;89(10):2124-2131.
8. Robinson CM, Howes J, Murdoch H, Will E, Graham C. Functional outcome and risk of recurrent instability after primary traumatic anterior shoulder dislocation in young patients. *J Bone Joint Surg Am.* 2006;88(11):2326-2336.
9. Bushnell BD, Creighton RA, Herring MM. Bony instability of the shoulder. *Arthroscopy.* 2008;24(9):1061-1073.

10. Beighton PH, Horan FT. Dominant inheritance in familial generalised articular hypermobility. *J Bone Joint Surg Br.* 1970;52:145-147.
11. Murrell GA, Warren RF. The surgical treatment of posterior shoulder instability. *Clin Sports Med.* 1995;14:903.
12. Harryman DT, Sidles JA, Harris SL, et al. The role of the rotator interval capsule in passive motion and stability of the shoulder. *J Bone Joint Surg Am.* 1992;74:53-66.
13. Blasier RB, Soslowsky LJ, Malicky DM, Palmer ML. Posterior glenohumeral subluxation: active and passive stabilization in a biomechanical model. *J Bone Joint Surg Am.* 1997;79A:433-440.
14. Kim SH, Ha KI, Park JH, et al. Arthroscopic posterior labral repair and capsular shift for traumatic unidirectional recurrent posterior subluxation of the shoulder. *J Bone Joint Surg Am.* 2003;85:1479-1487.
15. Altchek DW, Warren RF, Skyhar MJ, Ortiz G. T-Plasty modification of the Bankart procedure for multidirectional instability of the anterior and inferior types. *J Bone Joint Surg Am.* 1991;73A:105-112.
16. Bigliani LU, Kurzweil PR, Schwartzbach CC, Wolfe IN, Flatow EL. Inferior capsular shift procedure for anterior-inferior shoulder instability in athletes. *Am J Sports Med.* 1994;22:578-584.
17. Neer CS II, Foster CR. Inferior capsular shift for involuntary inferior and multidirectional instability of the shoulder: a preliminary report. *J Bone Joint Surg Am.* 1980;62A:897-908.
18. Cole BJ, L'Insalata J, Irrgang J, Warner JJP. Comparison of arthroscopic and open anterior shoulder stabilization: a two- to six-year follow up study. *J Bone Joint Surg Am.* 2000;82A:1108-1114.
19. Rook RT, Savoie FH III, Field LD. Arthroscopic treatment of instability attributable to capsular injury or laxity. *Clin Orthop.* 2001;390:52-58.
20. Bradley JP, Baker CL 3rd, Kline AJ, Armfield DR, Chhabra A. Arthroscopic capsulolabral reconstruction for posterior instability of the shoulder: a prospective study of 100 shoulders. *Am J Sports Med.* 2006;34(7):1061-1071.
21. Radkowski CA, Chhabra A, Baker CL 3rd, Tejwani SG, Bradley JP. Arthroscopic capsulolabral repair for posterior shoulder instability in throwing athletes compared with nonthrowing athletes. *Am J Sports Med.* 2008;36(4):693-699.
22. Caprise PA Jr, Sekiya JK. Open and arthroscopic treatment of multidirectional instability of the shoulder. *Arthroscopy.* 2006;22(10):1126-1131.
23. McIntyre LF, Caspari RB, Savoie FH III. The arthroscopic treatment of multidirectional instability: two-year results of a multiple suture technique. *Arthroscopy.* 1997;13:418-425.
24. Field LD, Warren RF, O'Brien SJ, Altchek DW, Wickiewicz TL. Isolated closure of rotator interval defects for shoulder instability. *Am J Sports Med.* 1995;23:557-563.
25. Gartsman GM, Taverna E, Hammerman SM. Arthroscopic rotator interval repair in glenohumeral instability: description of an operative technique. *Arthroscopy.* 1999;15:330-332.

4

SUPERIOR
LABRAL TEARS

Michael G. Ciccotti, MD

INTRODUCTION

The shoulder is exposed to a tremendous variety of forces through the full spectrum of sports, particularly overhead or throwing sports. Normally, there is a precise balance between the bony and soft tissues of the shoulder joint, which allows athletes to perform overhead sports smoothly and painlessly. Numerous studies have defined the various phases of throwing from windup through cocking, acceleration, deceleration, and follow-through (Figure 4-1).[1] During the various phases of throwing, significant forces are transmitted through the shoulder, specifically at the superior labrum, which can lead to injury.

Cohen SB. *Musculoskeletal Examination of the Shoulder:
Making the Complex Simple* (pp. 74-92).
© 2011 SLACK Incorporated

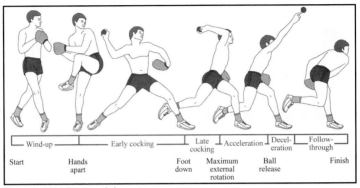

Figure 4-1. Phases of throwing.

Anatomically, the superior labrum is a fibrocartilaginous "bumper-like" structure that serves as an insertion site for the intra-articular portion of the biceps. Its anatomic form ranges from smooth attachment to the superior articular margin to a meniscoid-type structure extending over the articular edge.[2-4] Biomechanically, it serves to deepen the surface upon which the humeral head glides as well as serving as an attachment site for the biceps tendon and thereby transmitting significant force during the overhead throwing mechanism.[5,6]

HISTORY

Superior labral tears have most often been associated with chronic repetitive overhead activity, such as throwing; however, they also have been noted after single traumatic events, such as falling on an outstretched arm. Pain is most often described as deep or on the posterior aspect of the shoulder joint. In the throwing athlete, the pain may be slowly progressive or may be dramatically worsened with one throwing event. The patient may also complain of pain with overhead activity, difficulty sleeping, and a palpable and/or audible click or pop.

EXAMINATION

Physical examination of the overhead throwing athlete's shoulder is similar to that for other pathologies of the shoulder and includes a thorough inspection, palpation, evaluation of range of motion, and strength testing as well as stability testing (Table 4-1; see also Table 3-2). Inspection does not routinely identify any significant abnormalities, although the athlete may be noted to have greater muscle mass in the dominant shoulder and upper extremity as compared to the nondominant shoulder and upper extremity. The athlete's shoulder can also exhibit scapular dyskinesia as evidenced by elevation of the medial border of the scapula, scapular protraction, and scapular depression. Palpation may identify some pain along the posterior glenohumeral joint line distal and medial to the posterolateral corner of the acromion. Throwers with superior labral tears may also exhibit tenderness along the bicipital groove anteriorly. If they have any irritation or inflammation of the extra-articular portion of the biceps tendon, they may also have a positive Speed's and Yergason's test.

Range of motion testing is usually normal. It is important to note that the athlete's shoulder often has a shifted arc of motion compared to the contralateral shoulder. Throwers have increased external rotation and decreased internal rotation in the dominant shoulder; however, the overall arc of motion should be symmetric with the nondominant shoulder. Decreases in internal rotation can also be noted in the thrower, and if internal rotation deficits in the scapula plane are greater than 25 degrees or if the overall arc of motion is 25 degrees less than the contralateral shoulder, then glenohumeral internal rotation deficit (GIRD) may also be present (Figure 4-2).[7,8]

Strength testing is performed in order to assess the integrity of the rotator cuff and scapular stabilizing muscles as well as the core, trunk, and leg muscles. The kinetic chain theory[6,8-11] suggests that the force necessary for throwing is initiated in the lower extremities, transmitted through the pelvis, and continued by trunk rotation to the upper extremity through the shoulder joint. Weakness anywhere along this kinetic chain can lead to increased forces in the shoulder, resulting in a variety of pathologies, most notably superior labral injury.

Table 4-1

METHODS FOR EXAMINING THE SHOULDER WITH SUPERIOR LABRAL ANTERIOR POSTERIOR TEAR

Examination	Technique	Illustration	Grading	Significance
O'Brien test	1. Shoulder to 90 degrees flexion, 30 degrees of adduction, and thumb pointing downward with resisted forward flexion 2. Rotate forearm to full supination and resist forward flexion	 	Positive = pain or click	1. Pain or click worsened with thumb pointing downward compared to full supination = SLAP tear 2. Anterosuperior pain with thumb pointing downward and full supination = acromioclavicular joint pathology
Mayo shear test	Arm brought up into forward flexion, rotated into abduction		Positive = pain and/or clunk	Indicator of SLAP tear

(continued)

Table 4-1 (continued)

METHODS FOR EXAMINING THE SHOULDER WITH SUPERIOR LABRAL ANTERIOR POSTERIOR TEAR

Examination	Technique	Illustration	Grading	Significance
Speed's test	Shoulder flexion with elbow in full extension and forearm in full supination		Positive = pain	Indicates biceps tendonitis or SLAP tear
Yergason's test	Resisted forearm supination with the elbow extended		Positive = pain	Indicates biceps tendonitis or SLAP tear

(continued)

Table 4-1 (continued)

METHODS FOR EXAMINING THE SHOULDER WITH SUPERIOR LABRAL ANTERIOR POSTERIOR TEAR

Examination	Technique	Illustration	Grading	Significance
Scapular positioning	Observe medial border of scapula while patient slowly forward flexes shoulder 3 times	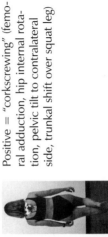	I = inferior angle elevated II = medial border elevated III = inferior angle, medial border and superior angle elevated	Weakness of periscapular stabilizing musculature
Core stability/flexibility testing	Sagittal, coronal, transverse planes		Inability to stabilize core over pelvis	Weakness of core/trunk musculature
Dynamic Trendelenberg test	Single-leg half-squat		Positive = "corkscrewing" (femoral adduction, hip internal rotation, pelvic tilt to contralateral side, trunkal shift over squat leg)	Weakness of lower extremity and pelvic musculature

(Images [except Mayo shear test] are reprinted with permission from Konin JG, Wiksten DL, Isear Jr JA, Brader H. Special Tests for Orthopedic Examination. 3rd ed. Thorofare, NJ: SLACK Incorporated; 2006.)

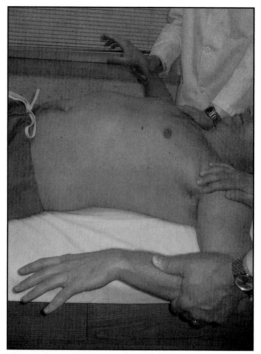

Figure 4-2. GIRD on right shoulder.

Lower extremity strength can be determined by the dynamic Trendelenburg test. The patient is asked to do a single-leg half-squat. Weakness of the lower extremity and pelvic musculature is indicated by "corkscrewing" (femoral adduction, hip internal rotation, pelvic tilt to the contralateral side, and trunkal shift over the squat leg).[12] Core stability testing of the trunk involves triplanar motion and flexibility and should be evaluated in the sagittal, coronal, and transverse planes. The scapula should also be assessed for both its static and dynamic positioning.[8] Elevation of its medial border suggests weakness of the periscapular stabilizing muscles.[8] Rotator cuff strength should be assessed by resisted internal and external rotation testing with the arm at the side, as well as by resisting forward elevation at 90 degrees in the scapular plane (Figure 4-3). Weakness and irritation of the rotator cuff may also be identified by the Neer and Hawkins impingement tests (Figures 4-4 and 4-5).

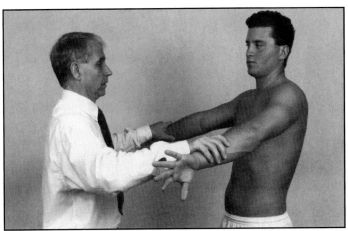

Figure 4-3. Rotator cuff testing. (Reprinted with permission from Konin JG, Wiksten DL, Isear Jr JA, Brader H. *Special Tests for Orthopedic Examination.* 3d ed. Thorofare, NJ: SLACK Incorporated; 2006.)

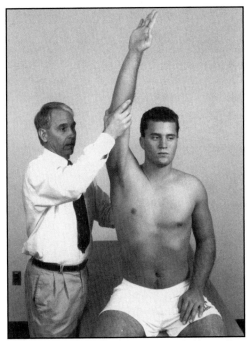

Figure 4-4. Neer test. (Reprinted with permission from Konin JG, Wiksten DL, Isear Jr JA, Brader H. *Special Tests for Orthopedic Examination.* 3d ed. Thorofare, NJ: SLACK Incorporated; 2006.)

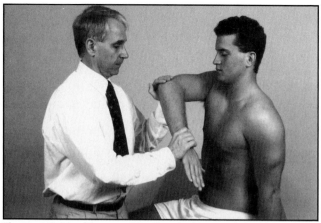

Figure 4-5. Hawkins test. (Reprinted with permission from Konin JG, Wiksten DL, Isear Jr JA, Brader H. *Special Tests for Orthopedic Examination.* 3d ed. Thorofare, NJ: SLACK Incorporated; 2006.)

Numerous specific tests have been suggested for identification of a superior labral tear (see Table 4-1). The active compression test (O'Brien test) and the Mayo shear test may be 2 of the more helpful clinical tests to identify superior labral anterior posterior (SLAP) lesions. The active compression test[11] is performed with the patient either sitting or in the supine position to stabilize the scapula. The shoulder is then placed in 90 degrees of abduction, the elbow extended, and the arm in 15 to 20 degrees of adduction. Resisted forward elevation with the thumb pointing downward causing pain deep or along the posterior shoulder and relieved with the thumb pointing upward is indicative of superior labral pathology. The Mayo shear test[13-15] is performed by circumducting the forward flexed shoulder. A painful click or pop deep or along the posterior joint line suggests a SLAP tear. Multiple studies, however, have shown that superior labral tears are clinically ambiguous and that no single test or combination of tests reliably predicts superior labral tears.[13-18] Burkhart[16] has correlated clinical testing with arthroscopically identified superior labral tears for type II SLAP lesions in an effort to more precisely determine clinical tests that may be helpful in diagnosing this entity. He noted that in those patients who had an identified type II SLAP tear extending from the base of the biceps anteriorly, preoperative symptoms included anterior bicipital groove tenderness and positive Speed's and O'Brien tests. He noted that in those patients who had an arthroscopically identified type II superior labral tear extending from the base of the biceps posteriorly, the apprehension/relocation test by Jobe was positive for posterior shoulder pain relieved with reduction of the posteriorly directed force. He noted that in those patients who had a type II SLAP tear extending across the entire superior labral from anterior to posterior, both anterior-type symptoms (bicipital groove tenderness, positive O'Brien test) as well as posterior symptoms (apprehension/relocation test for posterior pain) were present.

PATHOANATOMY

The precise mechanism of injury to the superior labrum has been debated. It may occur from an acute traumatic event while falling on an outstretched arm.[4] Alternatively, superior labral injuries may occur from the chronic, repetitive stress of overhead throwing. In throwing athletes, these injuries may be generated by 1 of 2 possible mechanisms. Several authors have suggested that biceps contraction during the follow through phase of throwing can generate a traction force on the superior labrum, leading to an avulsion-type injury.[6,19,20] Others have suggested that injury to the superior labrum occurs during the late cocking and early acceleration phases of throwing, where the humeral head rolls over the superior labrum, creating a "peel-back" type of injury to the superior labrum.[6,21,22] More recently, it has been suggested that superior labral injury in throwers may be caused by a combination of both of these mechanisms in a "weed pulling" manner.[6] It has been postulated that a predictable series of events may occur in the athlete's shoulder that lead to superior labral tears. This may begin with osseous adaptive changes involving a progressive humeral head retroversion. Scapular and rotator cuff muscular weakness may follow with repetitive throwing. Posterior soft tissue contracture may then occur due to the significant eccentric load forces that are generated in the posterior aspect of the shoulder during the deceleration follow-through phase of throwing. This combination then leads to a posterior superior migration of the humeral head on the glenoid with abduction and external rotation of the shoulder.

This may then generate the peel-back–type force that can lead to a superior labral injury as well as posterior superior rotator cuff injury (internal impingement).[7,9,10,] This cascade of events can occur in the thrower's shoulder but does not routinely occur in the general population.

Steve Snyder in 1990 initially described a classification system for injury to the superior labrum.[4] Other authors have since subclassified Snyder's type II superior labral tear into 3 subcategories.[23] Additionally, superior labral tears have been identified with a variety of other shoulder pathologies, including anterior labral tears, anterior capsular laxity, posterior labral tears, posterior capsular contracture, and rotator cuff tears (Table 4-2).

IMAGING

Routine plain x-ray views including an anteroposterior (AP) view of the shoulder joint in internal and external rotation, scapular Y view, and axillary view are most often normal in patients who have superior labral tears. Magnetic resonance imaging (MRI) with and without gadolinium arthrography enhancement has been suggested for identification of superior labral tears.[24,25] A meniscoid-like variant of the superior labrum has been identified in normal patients; however, separation of the labrum from the glenoid is rare posterior to the biceps root.[24,25] Also, paralabral cysts have a high correlation with superior labral tears. Contrast extravasation between the base of the superior labrum and the superior glenoid rim is consistent with a superior labral tear (Figure 4-6).

TREATMENT

Nonoperative

Superior labral injuries may be treated nonoperatively initially. This includes resting the patient from any provocative activities, especially overhead work or sport for 3 to 6 weeks depending upon the severity of pain. Heat and ice contrast as well as nonsteroidal anti-inflammatory drugs (NSAIDs) and

Table 4-2

HELPFUL HINTS

Type of Superior Labral Tear and Treatment

TYPE	DESCRIPTION	TREATMENT
I	Fraying; no detachment	Débridement
II	Detachment of biceps anchor from superior glenoid rim	Repair
A	Root of biceps and anterior	
B	Root of biceps and posterior	
C	Entire superior labrum	
III	Bucket handle tear; no detachment	Débridement of bucket portion
IV	Bucket handle tear with extension up into biceps tendon	Débridement of bucket portion and débridement/repair/tenodesis of biceps
V	Type II and anterior labral tear	Repair
VI	Flap tear of superior labrum	Débridement/repair
VII	Type II and extension into middle glenohumeral ligament (MGHL)	Repair SLAP; do not attach MGHL to anterior glenoid
VIII	Type II and posterior labral tear	Repair
IX	Circumferential labral tear	Repair
X	Type II and extension into rotator interval (RI)	Repair SLAP plus closure of RI

Imaging

IMAGING	PERTINENT VIEW	FINDINGS
Plain radiographs	AP in internal rotation	Most often normal in isolated SLAP tears
	AP in external rotation	
	Scapular—Y view	
	Axillary view	
MRI (arthrogram)	Axial, coronal, sagittal, ABER views	Evaluate for dye extravasation between the superior labrum and the superior glenoid rim with irregular undersurface of superior labrum ± paralabral cysts evaluate for concomitant rotator cuff fraying or tearing

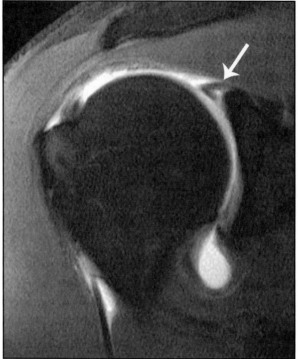

Figure 4-6. MRI of SLAP tear.

possibly an intra-articular injection of corticosteroid may be helpful to relieve the pain associated with the injury. Once pain has resolved, then a range of motion program including gentle passive, active assisted, and active range of motion begins. Particular attention is paid to the posterior shoulder soft tissues in order to resolve any posterior contracture or glenohumeral internal rotation deficit (Figure 4-7). Special attention is also directed to the scapula with respect to its position on the posterior chest wall. In addition to range of motion exercises, progressive strengthening exercises are carried out for the rotator cuff and scapular muscles. Special attention is directed to the scapular stabilizing muscles.[8,16] Any deficits in core, trunk, and lower extremity strength are addressed concomitantly. At 3 to 6 weeks, a progressive tossing/throwing program may be initiated. This begins with gentle tossing at a distance of 30 feet progressing sequentially to 180 feet.

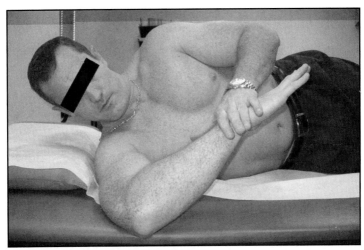

Figure 4-7. "Sleeper stretch" photo.

At that point, pitchers will then begin throwing from a mound, first with fast balls at increasing force followed by off-speed pitches. Return to competition with nonoperative treatment may be upwards of 6 to 12 weeks. If symptoms persist in spite of a structured nonoperative program, then operative treatment is considered.

Operative

Operative treatment is indicated in 1) throwing athletes who have persistent symptoms in spite of a well-coordinated nonoperative program; 2) elite-level athletes who have severe initial symptoms with clear-cut physical findings and easily identifiable labral tears on imaging studies; and 3) nonathletes who have persistent symptoms that interfere with daily activities, work, or sleep. The precise operative treatment is dependent upon the type of superior labral tear (see Table 4-2). Type I superior labral tears are most often débrided. Type II labral tears require repair. Type III labral tears may be treated by excision of the bucket handle-type portion with subsequent repair of the remaining superior labrum if it is unstable. Type IV superior labral tears may be treated by either excision of the bucket handle portion with extension into the biceps or repair of this portion. Treatment of types V through X is most often

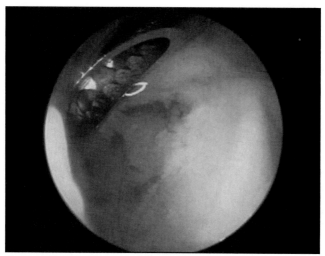

Figure 4-8. Débridement of fibrous tissue.

repair of the labral tear. The treatment of the biceps itself is dependent upon the degree of injury and ranges from débridement to repair of any torn portion to tenodesis. Limited data are available in the throwing athlete on tenodesis for severe intra-articular biceps tendon pathology. The most common type of superior labral tear requiring surgical treatment is type II.[4,16,26] The general principles of arthroscopic treatment include débridement of the frayed portion of the labrum, removal of any fibrous scar tissue, preparation of the superior glenoid rim to a healthy bleeding surface, and firm reattachment of the superior labrum to the superior glenoid rim bed. This can be performed either in a lateral decubitus or beach chair-type position and requires a viewing portal (most often posterior), an instrumentation and suture management portal (most often anterior), as well as either percutaneous placement or mid-lateral portal placement of implants and knot tying.[27] Our recommended treatment of choice includes lateral decubitus positioning with viewing from the posterior portal. The anterior portal is created in order to prepare the superior glenoid rim with a combination of rasp, motorized shaver, and motorized burr with care to protect the superior labrum while preparing the glenoid rim (Figure 4-8). This is done until a bleeding surface is identified under direct visualization (Figure 4-9).

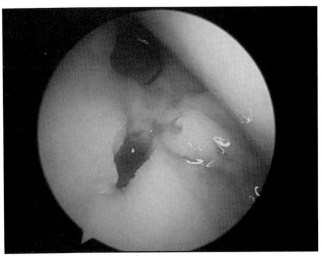

Figure 4-9. Prep of glenoid rim.

A midlateral portal is then created for implant placement. The precise position of the implants is determined by the type of tear. Implants are then placed under direct visualization into the superior glenoid rim. Sutures are then shuttled around the superior labrum, and knots are then tied sequentially, making certain that the knots are located on the nonarticular, peripheral portion of the labrum away from the articular margin and with care not to constrain the biceps root (Figure 4-10). Knotless and tack-type devices have also been proposed for superior labral repair. Other types of implants have advantages and disadvantages, but currently the most common type of implant used is a suture-tying implant. It is important to also address any concomitant pathology that may occur, including anterior labral fraying, tearing, and anterior capsular laxity as well as posterior labral fraying or tearing or posterior capsular contracture. If the rotator cuff is damaged, then appropriate treatment is necessary, ranging from débridement for minor cuff fraying to repair for near-complete or complete rotator cuff tears (see Chapter 6).[3,4,16]

Postoperative rehabilitation after superior labral repair includes sling use for 3 weeks. During this time, patients may use their hand, wrist, and elbow actively. They begin a gentle active-assisted range of motion program under the guidance

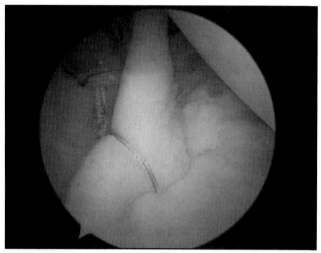

Figure 4-10. Repair of SLAP tear.

of a physical therapist or athletic trainer at approximately 2 weeks after surgery. Gentle strengthening begins between the second and fourth weeks. Progressive active-assisted and active range of motion then follows. Resistance-type strengthening begins at approximately 6 weeks. Overhead or throwing athletes begin a tossing program at 4 months after surgery, and it may be 9 to 12 months before overhead or throwing athletes can return to their preinjury level of competition. Nonathletes may return to preinjury activities at 4 to 6 months postoperatively. Results of arthroscopic superior labral repair suggest that approximately 80% to 90% of overhead athletes may return to the preinjury level of sport after successful repair.[8,16,19,26,28]

CONCLUSION

Anatomic form and biomechanical function are most closely associated in the superior labrum of the shoulder. Overhead athletes expose the superior labrum to tremendous forces during throwing. The precise mechanism of injury to the superior labrum may involve either a deceleration traction type of injury or a peel-back type of injury when the arm is in abduction and external rotation. The clinical diagnosis is extremely challenging with no specific tests as yet identified to precisely define superior labral tears. MRI with intra-articular gadolinium has helped greatly in identifying the presence and extent of superior labral tears. Nonoperative treatment may be carried out; however, most often, arthroscopic superior labral repair is necessary in the high-level or elite thrower. A structured postoperative rehabilitation program focusing on shoulder range of motion and strength; correcting any deficiencies in core, trunk, and lower extremity weakness; and improving poor mechanics of throwing is essential. It may take 4 to 6 months to return non-throwing patients to their pre-injury level of activity, and it may take even 9 to 12 months to return the throwing athlete to competition at his or her pre-injury level.

REFERENCES

1. Kvitne RS, Jobe FW. The diagnosis and treatment of anterior instability in the throwing athlete. *Clin Orthop Relat Res.* 1993;291:107-123.
2. Davidson AE. Mobile superior glenoid labrum: a normal variant or pathologic condition? *Am J Sports Med.* 2004;32(4):962-966.
3. Ilahi OF, Labbe MR, Cosculluela P. Variants of the anterosuperior labrum and associated pathology. *Arthroscopy.* 2002;18(8):882-886.
4. Snyder SJ, Karzell PP, Delpizzo W, et al. SLAP lesions of the shoulder. *Arthroscopy.* 1990;6:274-279.
5. Andrews JR, Carson WG, McCleod WD. Glenoid labrum tears related to the long head of the biceps. *Am J Sports Med.* 1985;13:337-341.
6. Kibler WB. Biomechanical analysis of shoulder during tennis activities. *Clin Sports Med.* 1995;14:79-86.
7. Kibler WB. The role of the scapula in the athletic shoulder function. *Am J Sports Med.* 1998;26:325-337.
8. Kibler WB, Livingston B, Chandler TJ. Shoulder rehabilitation: clinical application, evaluation, and rehabilitation protocols. *Instr Course Lect.* 1997;46:43-52.
9. Myers JB, Laudner KG, Pasquale MR, Bradley JP, Lephart SM. Glenohumeral range of motion deficits and posterior shoulder tightness in throwers with pathologic internal impingement. *Am J Sports Med.* 2006;34(3):385-391.
10. Hirashima M, Yamane K, Nakamura Y, Ohtsuki T. Kinetic chain of overarm throwing in terms of joint rotations revealed by induced acceleration analysis. *J Biomech.* 2008;41(13):2874-2883.
11. O'Brien SJ, Panani MJ, McGlynn SR. The active compression test. *Am J Sports Med.* 1998;26:610-613.

12. Burkhart SS, Morgan CD, Kibler WB. The disabled throwing shoulder: spectrum of pathology part III: the SICK scapula, scapular dyskinesia, the kinetic chain, and rehabilitation. *Arthroscopy.* 2003;19(6):641-661.
13. Burkhart SS. Shoulder injuries in the overhead athlete: the dead arm revisited. *Clin Sports Med.* 2000;19:125-158.
14. Kim TK. Clinical features of the different types of SLAP lesions. *J Bone Joint Surg.* 2006;85-A:66-71.
15. McFarland EG. *Examination of the Shoulder.* New York, NY: Thieme; 2006.
16. Burkhart SS, Morgan CD, Kibler WB. The disabled throwing shoulder: spectrum of pathology part II: evaluation and treatment on SLAP lesions in throwers. *Arthroscopy.* 2003;19(5):531-539.
17. Stetson WB, Templin K. The crank test, the O'Brien test, and routine magnetic resonance imaging scans in the diagnosis of labral tears. *Am J Sports Med.* 2002;30(6):806-809.
18. Parentis MA, Glousman RE, Mohr KS, Yocum LA. An evaluation of the provocative tests for superior labral anterior posterior lesions. *Am J Sports Med.* 2006;34(2):265-268.
19. Maffet MW, Gartsman GM, Moseley B. Superior labrum-biceps tendon complex lesions of the shoulder. *Am J Sports Med.* 1995;23:93-98.
20. Burkhart SS, Morgan CD. Technical note: the peel-back mechanism: its role in producing and extending posterior type II SLAP lesions and its effect on SLAP repair rehabilitation. *Arthroscopy.* 1998;14:637-640.
21. Shepard MF, Dugas JR, Zeng N, Andrews JR. Differences in the ultimate strength of the biceps anchor and the generation of type II superior labral anterior posterior lesions in a cadaveric model. *Am J Sports Med.* 2004;32:1197-1201.
22. Grossman MG, Tibone JE, McGarry MH. A cadaveric model of the shoulder: a possible etiology of superior labrum anterior-to-posterior lesions. *J Bone Joint Surg.* 2005;87:824-831.
23. Morgan CD, Burkhart SS, Palmeri M,Gillespie M. Type II SLAP lesions: three subtypes and their relationships to superior instability and rotator cuff tears. *Arthroscopy.* 1998;14:553-565.
24. Mohana-Borges AV, Chung CB, Resnick D. Superior labral anteroposterior tear: classification and diagnosis of MRI and MR arthrography. *Am J Roentgenol.* 2003;181:1449-1462.
25. Neumann CH. MR imaging of the labral-capsular complex: normal variants. *Am J Roentgenol.* 1991;157:1015-1021.
26. Limpisvasti O, El Attrache NS. Understanding shoulder and elbow injuries in baseball. *JAAOS.* 2007;15:139-147.
27. Ciccotti MG, Kuri J, Leland JM, Schwartz M, Becker C. A cadaveric analysis of the arthroscopic fixation of anterior and posterior SLAP lesions through a novel lateral transmuscular portal. *Arthroscopy.* 2010;26(1):12-18.
28. Neuman B, Boisvert B, Ciccotti MG, et al. Results of arthroscopic repair of type II SLAP lesions in overhead athletes: assessment of return to pre-injury playing level and satisfaction. Presented at: American Orthopaedic Society for Sports Medicine 36th Annual Meeting; July 15-18, 2010; Providence, RI.

5

ROTATOR CUFF

Gregg J. Jarit, MD and David R. Diduch, MD

INTRODUCTION

Rotator cuff disorders affect a large percentage of the general population and can prevent patients from performing normal basic activities of daily living.[1] They can be very debilitating and will often prevent the patient from sleeping. This condition can affect manual laborers, athletes, or sedentary individuals. Advances in the understanding of the pathoanatomy of the rotator cuff and in the arthroscopic treatment of these disorders have increased the success rate of returning patients to their chosen activities.

Cohen SB. *Musculoskeletal Exam of the Shoulder:*
Making the Complex Simple (pp. 93-111).
© 2011 SLACK Incorporated

HISTORY

While the history can be varied in patients with rotator cuff tears, a common presentation is in a middle-aged patient who experiences a fall either onto his or her outstretched hand or directly onto his or her shoulder. Patients can describe their pain as anterior, lateral, or posterior in the shoulder, and the pain often radiates down the anterior or lateral aspect of the arm but not below the elbow. There will also be a subset of patients who do not describe a traumatic event. These are often patients who have performed overhead activities in sports, work, or at home and have developed pain with certain arm positions, especially overhead. The symptoms may be insidious over several months. In older patients who sustain a traumatic shoulder dislocation, a rotator cuff tear is extremely common and may be seen in up to 100% of patients older than 70 years.[2]

An important question to ask the patient with shoulder pain in the setting of suspected rotator cuff pathology is whether the pain wakes him or her up at night or prevents him or her from sleeping.[3] Patients will often describe being awakened by pain if they have inadvertently rolled onto the affected shoulder, or they will wake up on their back after having gone to sleep on the affected side. The patient will often have tried pain medication such as acetaminophen or nonsteroidal anti-inflammatory drugs (NSAIDs). These medications may provide mild relief but are less effective in the presence of a symptomatic complete rotator cuff tear. The same is true for physical therapy, which can be very helpful for subacromial impingement and with partial rotator cuff tears (less than 50%) but may be less reliable for complete tears.

EXAMINATION

The examination starts with observation and inspection of the skin and bony structures of the shoulder. In a patient with longstanding rotator cuff disease, there may be generalized muscular atrophy on the affected side. There may also be atrophy visible posteriorly above or below the scapular spine, corresponding to the supraspinatous and infraspinatous,

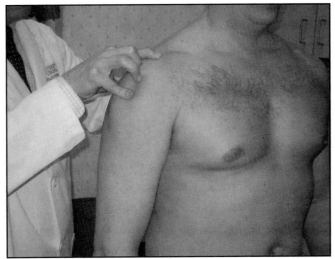

Figure 5-1. Palpation of the AC joint.

respectively. The next part of the examination is palpation. The examiner should palpate the bony shoulder girdle, including the clavicle, acromioclavicular (AC) joint (Figure 5-1), acromion, bicipital groove, and scapular spine. Patients with rotator cuff disease often have concomitant shoulder pathology, such as AC joint osteoarthritis, and these conditions need to be identified as well. Treating the rotator cuff alone in the setting of other conditions may not completely alleviate the patient's shoulder pain. The long head of the biceps tendon is often involved as well and can be palpated in the bicipital groove (Figure 5-2).

The next area to be assessed is range of motion (ROM), including forward elevation, abduction, external rotation, and internal rotation. These values should be compared to the contralateral side to assess for loss of motion. In general, the patient should be able to actively forward elevate and abduct to 180 degrees, externally rotate at the side to 50 to 60 degrees, and internally rotate at the side to the lower thoracic spine. In patients with rotator cuff tears, their loss of ROM will depend on which part of the rotator cuff is affected and how much compensation they are receiving from other muscles around the shoulder girdle. Patients with supraspinatous

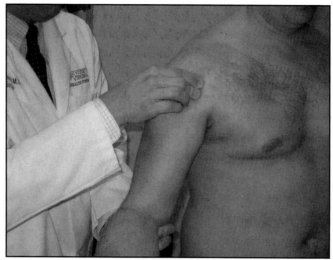

Figure 5-2. Palpation of the bicipital groove.

tears may have loss of forward elevation and abduction, while patients with infraspinatous tears may have loss of active external rotation. If there is a loss of motion in any of these planes, it is important to also test the passive ROM. Patients with rotator cuff tears should have greater passive than active ROM. If they do not, adhesive capsulitis should be considered. Some stiffness can be expected, however, with chronic rotator cuff tears.

The next aspect of the physical exam includes strength and provocative testing of the rotator cuff (Table 5-1). Often, strength deficits are the main indicator that a full-thickness rotator cuff tear is present, as partial-thickness tears may produce more pain than weakness compared to full tears.[4] External rotation strength with the arm at the side should be tested. Significant weakness with this test is sensitive for a large tear and is thus a good screening test. This is followed by abduction strength in the scapular plane. This is approximately 30 degrees anterior to the coronal line through the body, and the patient's thumbs should be directed toward the floor, which isolates the supraspinatous muscle (empty can test). This test can then be repeated with the thumbs pointing up, which is more specific for the infraspinatous muscle.

Table 5-1

METHODS FOR EXAMINING THE ROTATOR CUFF

Examination	Technique	Illustration	Significance
Range of motion	Examine active and passive motion		Loss of active motion only may indicate dysfunction of rotator cuff
Strength testing	Manual strength testing: External rotation, abduction in scapular plane ("empty can," supraspinatous stress)		Loss of strength may indicate rotator cuff pathology

(continued)

Table 5-1 (continued)

METHODS FOR EXAMINING THE ROTATOR CUFF

Examination	Technique	Illustration	Significance
Belly press test	Patient presses against abdomen. Positive if patient brings elbow in toward the body		Subscapularis insufficiency
Lift-off test	Patient places the arm behind the back and tries to lift it away from the body. Positive if patient cannot perform		Subscapularis insufficiency
Outlet impinge-ment signs	Neer: Passive forward flex-ion of internally rotated arm Hawkins: Internal rotation of forward flexed arm		Both maneuvers place greater tuberosity in proximity to acro-mion. Pain present with external impingement and rotator cuff tears

(continued)

Table 5-1 (continued)

METHODS FOR EXAMINING THE ROTATOR CUFF

Examination	Technique	Illustration	Significance
Glenohumeral internal rotation heficit (GIRD)	Patient supine: Internal rotation of abducted shoulders		Deficit of internal rotation may indicate internal impingement but is nonspecific
Speed's test	Resisted shoulder flexion with elbow extended and forearm supinated		Positive with biceps tendon (long head) pathology, which is often concomitant with rotator cuff pathology
Yergason's test	Resisted supination of pronated forearm with elbow flexed to 90 degrees		Positive with biceps tendon (long head) pathology, which is often concomitant with rotator cuff pathology

With rotator cuff disease, patients will often have increased pain with abduction from 60 to 120 degrees. These strength tests will often elicit significant pain in the patient with rotator cuff tears but not as much in the patient with external subacromial impingement alone. The subscapularis muscle should be tested with the belly press test, "lift-off" test, or bear hug test. The belly press test is preferred if the patient has stiffness, especially in internal rotation, which is common. Significant external rotation weakness can be tested for with the external rotation lag test (Figure 5-3), in which the arm is passively rotated externally and let go. If the patient cannot keep his or her arm externally rotated, the test is considered positive.

Provocative testing continues with the Hawkins test for outlet impingement, which involves forward flexion to above 90 degrees, elbow flexion to 90 degrees, followed by passive internal rotation of the arm. This will be positive in patients with outlet impingement and rotator cuff tears. The same is true for the Neer test, which involves maximal forward flexion with the arm in internal rotation. Both maneuvers cause the greater tuberosity to contact or come close to the acromion, which would pinch any remaining rotator cuff tissue or the bursa and cause pain. Biceps tendon (long head) involvement should be tested with the Speed's test, which involves resisted forward elevation of the arm with the elbow slightly flexed, and Yergason's test, which involves resisted supination of the forearm from a pronated position and the elbow flexed to 90 degrees.

Another form of rotator cuff pathology is internal impingement, which is seen in overhead athletes such as baseball or volleyball players. In this condition, the patient develops glenohumeral internal rotation deficit (GIRD) first, with tightness of the posterior capsule. This can be tested by having the patient lie supine, abduct both arms to 90 degrees, and allow them to internally rotate (Figure 5-4). The affected side will have 20 to 30 degrees less internal rotation than the nonaffected side. This adversely affects the mechanics of the glenohumeral joint by causing a relative anterior translation of the humeral head. Subsequently, the articular side of the rotator cuff, in particular the infraspinatous, gets pinched with the arm in abduction and external rotation (the cocking position for throwers). Patients will have pain in this position as well

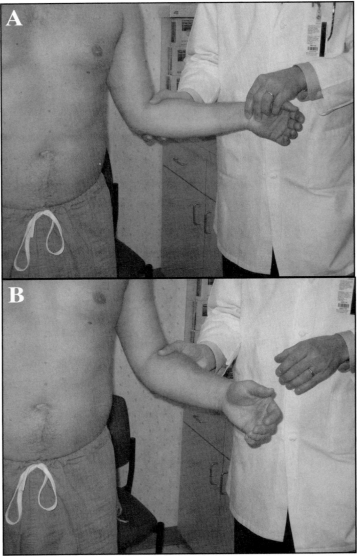

Figure 5-3. The external rotation lag test for integrity of the external rotators. (A) The arm is passively externally rotated. (B) The arm is released. The test is positive if the patient cannot hold the arm externally rotated.

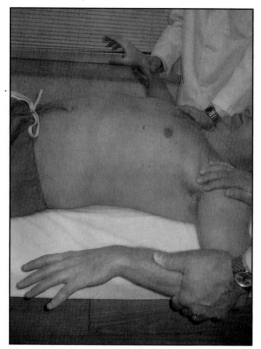

Figure 5-4. Test for evaluation of GIRD.

as during follow through, when the posterior capsule gets stretched. If untreated, this can also lead to "peel back" of the biceps tendon and a superior labral tear (see Chapter 5).

PATHOANATOMY

The rotator cuff consists of 4 tendons: subscapularis, supraspinatous, infraspinatous, and teres minor. The subscapularis originates on the ventral aspect of the scapula and inserts on the lesser tuberosity. It functions as a humeral head internal rotator. The supraspinatous originates in the suprascapular fossa and inserts onto the greater tuberosity and serves to abduct the arm. In particular, it helps to initiate abduction and also is the primary abductor in the scapular plane of motion. The infraspinatous originates in the infrascapular fossa posteriorly and inserts onto the posteroinferior portion of the greater tuberosity. It externally rotates the arm. The teres minor originates on the lateral border of the scapula and inserts onto the posterior aspect of the proximal humerus. It functions in external rotation at the side. Another main function of the rotator cuff is a dynamic stabilizer of the humeral head, keeping it centered in the glenoid throughout a functional ROM.

While any of these tendons can be torn, the most commonly torn tendon is the supraspinatous.[5] The etiology of rotator cuff tears is controversial and likely due to multiple factors. Anatomic factors that may contribute to both articular- and bursal-sided tears in this tendon involve the vascularity of the rotator cuff. There is a relative avascular zone at the supraspinatous insertion that leaves this area vulnerable to injury.[6,7] Injury can occur to the rotator cuff with external or outlet impingement, in which the greater tuberosity nears or comes into contact with the undersurface of the acromion (Table 5-2).[8] This can occur in an acute traumatic event or chronically with repetitive overhead motions. Some patients are predisposed to this type of problem based on the morphology of the acromion. Three types were originally described—type I is a flat acromion, type II is curved anteriorly, and type III has a hook anteriorly.[9] There is a higher incidence of external impingement with a type III acromion. A type IV acromion, with a convex undersurface near its distal end, was subsequently described but has not been shown to have an effect on the incidence of rotator cuff pathology.[10]

Another theory is that rotator cuff tears initiate on the articular side of the rotator cuff. With this mechanism, the fibers tear off of the footprint. It is possible that internal impingement can play a role in this type of tearing. With internal impingement, the posterosuperior rotator cuff is pinched between the humeral head and the glenoid labrum when the arm is in an abducted and externally rotated position. Another mechanism of tearing is with glenohumeral dislocation in a patient older than 40 years in which the proposed etiology is traction on a rotator cuff that is less compliant and flexible than in a younger population. The incidence of rotator cuff tear with glenohumeral dislocation in this population is very high, and the tear pattern tends to be complete avulsion of the tendon off of the greater tuberosity.[2] The lack of compliance and decreased flexibility likely plays a role in all etiologies of rotator cuff tearing.

Table 5-2

HELPFUL HINTS

Injury	Description of Injury	Typical Patient Complaint
External outlet impingement	Rotator cuff pinched under acromion with overhead positions	Pain with overhead and behind the back activities Strength generally preserved
Partial rotator cuff tear	Rotator cuff torn but some fibers remain attached throughout footprint	As above with night pain, especially when lying on affected side ± weakness
Full-thickness cuff tear	Full avulsion of rotator cuff attachment from footprint	As above with definite weakness

Imaging

IMAGE	PERTINENT IMAGE VIEWS	FINDINGS
Plain radiograph	AP, outlet views	Decreased subacromial distance Cyst formation and subchondral sclerosis at greater tuberosity
MRI (± arthrogram)	Coronal, sagittal	Discontinuity of rotator cuff tendons
CT—arthrogram	Coronal, sagittal	Discontinuity of rotator cuff tendons

Once the rotator cuff is torn, patients can experience significant pain with overhead activities due to external impingement on a torn area with exposed footprint or with any motion that causes traction on the torn tendon. In addition, the tear can continue to progress over time. While interstitial rotator cuff tears may be able to heal, tears where the tendon has avulsed from the footprint will not heal without surgical intervention. If not treated, the tear may continue to retract, get larger, and progress to a point that it is irreparable. A chronic retracted tendon sometimes cannot be mobilized enough to pull it back onto the footprint during surgery.

IMAGING

Imaging for suspected rotator cuff pathology (see Table 5-1) includes plain radiographs of the involved shoulder, including true anteroposterior (AP), scapular Y or outlet, Zanca (AC joint), and axillary views. Radiographs in patients with a rotator cuff tear may range from being normal to having a high-riding humeral head with complete obliteration of the subacromial space. The latter finding occurs in patients with longstanding massive rotator cuff tears that are usually not amenable to repair (rotator cuff arthropathy). Less than 6 mm of acromiohumeral distance on the AP view is considered narrowed. Most often, there may be more subtle radiographic findings, such as calcification in the area of the rotator cuff tendon (calcific tendonitis) or sclerotic changes and cyst formation at the greater tuberosity. The shape and type of the acromion can be evaluated on the scapular Y view, which may play a role in rotator cuff pathology and affect surgical management. The presence and degree of osteoarthritis in the glenohumeral and AC joints may also be assessed.

The gold standard for diagnosis of rotator cuff tears is magnetic resonance imaging (MRI).[11] This allows evaluation of the soft tissues around the shoulder and has been shown to be more than 90% sensitive and specific in the diagnosis of tears.[12,13] The accuracy of MRI may be further improved with the use of an arthrogram.[14] In addition to showing the presence of a full-thickness tear, an MRI can often determine a high-grade or low-grade partial thickness tear (Figure 5-5). For patients who are unable to undergo an MRI, a computed tomography (CT) arthrogram may also be performed (Figure 5-6). An ultrasound can also be obtained as an alternative; however, this requires a radiologist who is trained in the performance and interpretation of that procedure.

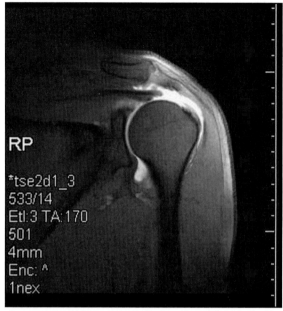

Figure 5-5. Oblique coronal MRI image demonstrating a full-thickness tear of the supraspinatous tendon with retraction.

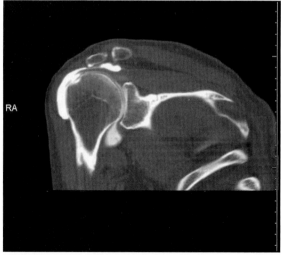

Figure 5-6. CT arthrogram image demonstrating a full-thickness tear of the supraspinatous tendon without significant retraction.

Table 5-3

TREATMENT OF ROTATOR CUFF PATHOLOGY IN SYMPTOMATIC PATIENTS

Diagnosis	Treatment
External outlet impingement	Rest, NSAIDs, physical therapy, possible injection
	Arthroscopic subacromial decompression if conservative treatment fails
	Most improve without surgery
Internal impingement	Sleeper stretch or correction of other underlying pathology
Partial rotator cuff tear (<50%)	Rest, NSAIDs, physical therapy, possible injection
	Arthroscopic subacromial decompression and débridement if conservative treatment fails
Partial rotator cuff tear (>50%)	Arthroscopic rotator cuff repair, ± subacromial decompression
Full-thickness rotator cuff tear	Arthroscopic rotator cuff repair ± subacromial decompression

TREATMENT

Patients with rotator cuff tendonitis or minor partial-thickness rotator cuff tears should be initially treated nonoperatively, starting with a course of oral NSAIDs or a subacromial corticosteroid injection (Table 5-3). If there is a specific offending activity or sport, it should be temporarily discontinued. A physical therapy protocol should be initiated with focus on rotator cuff strengthening and should address any ROM deficits. The emphasis for strengthening should be on the external rotators, with avoidance of offending arm positions, usually including keeping the arm below 90 degrees abduction. A subacromial injection of corticosteroid mixed with a local anesthetic is often performed to reduce acute inflammation and assist with the progress of physical therapy; however, the long-term benefit of these injections has been questioned.[15] Another group of patients who benefit from nonoperative management are those with GIRD. These patients respond very well to performance of the sleeper stretch, in which the patient lies directly on the affected side with the arm abducted to 90 degrees and passively stretches the arm into internal rotation.

Surgery can be considered in patients with full-thickness rotator cuff tears, partial-thickness tears, or recurrent subacromial impingement who have failed a course of nonoperative treatment. The treatment of choice is shoulder arthroscopy with débridement of the tear and subacromial decompression with rotator cuff repair. This has been shown to have improved results over open repair.[16,17] A diagnostic arthroscopy is first performed, assessing first the glenohumeral joint, including the cartilage surface, glenoid labrum, and biceps tendon. If biceps tendon pathology is found, a biceps tenotomy or tenodesis can be performed. The undersurface of the rotator cuff can be visualized as well. Placement of a "marker stitch" through a partial tear aids in identification of the tear from both surfaces and can be especially helpful in decision making for repair or débridement.

After intra-articular pathology is addressed, the subacromial space is inspected. This requires performance of a limited bursectomy. A subacromial decompression can be performed, especially in the presence of clear external impingement, bone spur formation on the undersurface of the acromion, or types III and IV acromial morphology. Patients with symptomatic AC joint arthritis can undergo a distal clavicle resection arthroscopically.

The decision to proceed with rotator cuff repair depends on the extent of the tear. Tears involving less than 50% of the thickness of the cuff are usually débrided. Along with subacromial decompression, this is often successful in alleviating symptoms.[18,19] All full-thickness tears and tears involving more than 50% of the cuff should be repaired, if possible.

Arthroscopic rotator cuff repair has significantly improved over the past several years and has become the favored technique of rotator cuff repair among many orthopedic surgeons.[20] There are unfortunately some tears that are too large and retracted to be repaired. In this case, the performance of débridement and limited decompression can provide some pain relief.[21] It is especially important in this setting to preserve the coracoacromial ligament, as this is often the only restraint to anterosuperior escape of the humeral head in the setting of massive irreparable rotator cuff tears.[22]

The postoperative course is dependent on whether the rotator cuff was repaired. If it was not, a sling can be used for comfort for the first few days after surgery, but should be removed frequently for immediate ROM exercises, often starting in the recovery room. Rotator cuff strengthening should begin as well and be completed before full return to activities or sports. If a repair is performed, the patient is placed in a sling postoperatively. While surgeons differ on the details of the postoperative protocol, use of the sling usually lasts 4 to 6 weeks. During that time, the sling may be removed for pendulum exercises and elbow, wrist, and hand ROM. Some physicians begin early passive forward elevation as well. After 4 to 6 weeks, the patient progresses to full active and passive ROM, followed by strengthening exercises. Full recovery can take up to 6 months.

CONCLUSION

Rotator cuff tears are debilitating injuries that affect a large percentage of the population. A careful history and physical examination with appropriate imaging studies are important for accurate diagnosis. The first line of treatment is usually nonoperative with rest, anti-inflammatory medication, physical therapy, and injections. Patients with symptomatic full-thickness tears or those who have failed nonoperative management are candidates for surgical intervention, usually arthroscopic repair.

REFERENCES

1. Yamaguchi K, Ditsios K, Middleton WD, et al. The demographic and morphological features of rotator cuff disease: a comparison of asymptomatic and symptomatic shoulders. *J Bone Joint Surg Am.* 2006;88(8):1699-1704.
2. Simank HG, Dauer G, Schneider S, Loew M. Incidence of rotator cuff tears in shoulder dislocations and results of therapy in older patients. *Arch Orthop Trauma Surg.* 2006;126(4):235-240.
3. Litaker D, Pioro M, El Bilbeisi H, Brems J. Returning to the bedside: using the history and physical examination to identify rotator cuff tears. *J Am Geriatr Soc.* 2000;48(12):1633-1637.
4. Fukuda H. Partial-thickness rotator cuff tears: a modern view on Codman's classic. *J Shoulder Elbow Surg.* 2000;9(2):163-168.
5. Cofield RH. Rotator cuff disease of the shoulder. *J Bone Joint Surg Am.* 1985;67(6):974-979.
6. Rothman RH, Parke WW. The vascular anatomy of the rotator cuff. *Clin Orthop Relat Res.* 1965;41:176-186.
7. Rathbun JB, Macnab I. The microvascular pattern of the rotator cuff. *J Bone Joint Surg Br.* 1970;52(3):540-553.
8. Neer CS II. Impingement lesions. *Clin Orthop Relat Res.* 1983;173:70-77.
9. Bigliani LU, April EW. The morphology of the acromion and its relationship to rotator cuff tears. *Orthopedic Transactions.* 1986;10:228.
10. Farley TE, Neumann CH, Steinbach LS, Petersen SA. The coracoacromial arch: MR evaluation and correlation with rotator cuff pathology. *Skeletal Radiol.* 1994;23(8):641-645.
11. Burk DL Jr, Karasick D, Kurtz AB, et al. Rotator cuff tears: prospective comparison of MR imaging with arthrography, sonography, and surgery. *Am J Roentgenol.* 1989;153(1):87-92.
12. Iannotti JP, Zlatkin MB, Esterhai JL, et al. Magnetic resonance imaging of the shoulder. Sensitivity, specificity, and predictive value. *J Bone Joint Surg Am.* 1991;73(1):17-29.
13. Zlatkin MB, Iannotti JP, Roberts MC, et al. Rotator cuff tears: diagnostic performance of MR imaging. *Radiology.* 1989;172(1):223-229.
14. Waldt S, Bruegel M, Mueller D, et al. Rotator cuff tears: assessment with MR arthrography in 275 patients with arthroscopic correlation. *Eur Radiol.* 2007;17(2):491-498.
15. Alvarez CM, Litchfield R, Jackowski D, Griffin S, Kirkley A. A prospective, double-blind, randomized clinical trial comparing subacromial injection of betamethasone and xylocaine to xylocaine alone in chronic rotator cuff tendinosis. *Am J Sports Med.* 2005;33(2):255-262.
16. Warner JJ, Tétreault P, Lehtinen J, Zurakowski D. Arthroscopic versus mini-open rotator cuff repair: a cohort comparison study. *Arthroscopy.* 2005;21(3):328-332.
17. Ide J, Maeda S, Takagi K. A comparison of arthroscopic and open rotator cuff repair. *Arthroscopy.* 2005;21(9):1090-1098.

18. Budoff JE, Nirschl RP, Guidi EJ. Débridement of partial-thickness tears of the rotator cuff without acromioplasty: long-term follow-up and review of the literature. *J Bone Joint Surg Am.* 1998;80(5):733-748.

19. Ogilvie-Harris DJ, Wiley AM. Arthroscopic surgery of the shoulder: a general appraisal. *J Bone Joint Surg Br.* 1986;68(2):201-207.

20. Youm T, Murray DH, Kubiak EN, Rokito AS, Zuckerman JD. Arthroscopic versus mini-open rotator cuff repair: a comparison of clinical outcomes and patient satisfaction. *J Shoulder Elbow Surg.* 2005;14(5):455-459.

21. Liem D, Lengers N, Dedy N, et al. Arthroscopic débridement of massive irreparable rotator cuff tears. *Arthroscopy.* 2008;24(7):743-748.

22. Wiley AM. Superior humeral dislocation: a complication following decompression and débridement for rotator cuff tears. *Clin Orthop Relat Res.* 1991;(263):135-141.

6

BICEPS TENDON

James R. Romanowski, MD and Mark W. Rodosky, MD

INTRODUCTION

Despite the relatively minor role in glenohumeral function, the long head of the biceps tendon (LHBT) is often responsible for significant shoulder disability. Anatomic constraints and various functional stresses experienced through a wide range of motion create an environment that makes the LHBT susceptible to injury. The morbidity associated with the proximal biceps is also related to its rich innervation and sensitivity to various biochemicals known to interact with the nervous system such as substance P and calcitonin gene-related peptides.[1]

Cohen SB. *Musculoskeletal Examination of the Shoulder:*
Making the Complex Simple (pp. 112-135).
© 2011 SLACK Incorporated

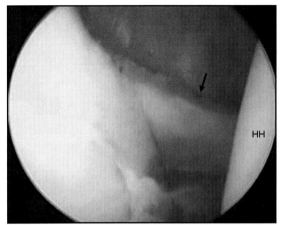

Figure 6-1. Arthroscopic view of biceps synovitis. View from posterior portal demonstrating early disease and inflammation (black arrow) of the LHBT. HH = humeral head.

Given the anatomic course of the biceps tendon, it is not surprising that it is rarely an isolated pathology, but is often associated with superior labral tear from anterior to posterior (SLAP) lesions, rotator cuff tears, biceps instability, and subacromial impingement (Figure 6-1). Primary biceps tendonitis is thought to comprise only 5% of the total cases of biceps tendonitis. The remaining 95% are related to these secondary causes.[2]

There is little debate over the function of the distal biceps tendon—elbow flexion and forearm supination. Proximally, however, there is little agreement or understanding on the major role of the long head. With external rotation of the humerus, the biceps has been shown to be a weak abductor.[3] There is also some evidence that the LHBT helps prevent proximal migration of the humeral head at the glenohumeral joint serving as a humeral head depressor.[4] For the overhead thrower, the LHBT contributes to anterior stability of the glenohumeral joint by preventing translation when activated.[5,6]

HISTORY

As with any patient evaluation, a detailed history is necessary to help identify the underlying causes for proximal biceps pathology and to develop a treatment plan in line with patients' functional demands and expectations. Acute trauma is a rare cause of proximal biceps tears but may result in biceps-related pain in situations where the tendon becomes unstable—subscapularis tears and SLAP lesions. "Clicking" or "snapping" with rotational overhead activities may point toward subluxation or dislocation of the biceps. A history of prior steroid injections, particularly into the bicipital tunnel, should be clarified, as this may suggest an additional iatrogenic source. Patients typically complain of anterior shoulder pain centered over the bicipital groove. Pain may be referred from other areas, especially in the setting of additional shoulder conditions. Radicular symptoms should be thoroughly evaluated, as this may be a significant source of referred pain. Fatigue, muscle cramping, "Popeye" deformity, and a "pop" are complaints related to attritional or acute ruptures of the tendon. The discomfort may be temporal, especially at night. Particular positions or activity can elicit the pain, such as overhead activity or motions that compress the biceps tendon under the coracoacromial arch, including abduction with rotation of the proximal humerus.

Because the majority of cases are secondary to other shoulder conditions, additional complaints are the norm and are related to those deficiencies. Given these overlapping and concomitant diagnoses, a thorough physical exam is critical for identification of patients' pathology.

EXAMINATION

The patient is observed in clothing that allows for simultaneous visualization of the bilateral upper extremities. Skin discoloration, ecchymosis, prior surgical incisions, atrophy, and/or tenting as a result of underlying bony deformities are noted (Table 6-1). Palpation is performed with a focus on the bicipital groove, which has a sensitivity of 53% and specificity of 54% for identifying pain generated by the LHBT (Table 6-2).[7]

Table 6-1

HELPFUL HINTS

Condition	Complaints	Examination Findings
Biceps tendonitis	Anterior pain in groove	Tenderness to palpation, positive Speed's and Yergason's tests
Proximal biceps rupture	"Popeye" deformity, cramping, fatigue, possible overhead activity pain	Shortened biceps muscle, mild weakness
Biceps tendon subluxation/ dislocation	Popping, internal rotation weakness	Positive biceps Instability test, weakness with internal rotation testing, positive belly press, lift-off, and bear hug tests

Exam	Technique	Pathology
Palpation	Direct pressure over biceps tendon and biceps groove	Biceps tendonitis
Speed's test	Shoulder forward flexion to 60 degrees and elbow flexion to 20 degrees. Patient resists downward force	Biceps tendonitis Superior labral pathology
Yergason's test	Elbow flexed to 90 degrees forearm pronated. Patient is asked to actively flex the elbow and supinate the wrist against resistance. Pain is referred to bicipital groove	Biceps tendonitis Superior labral pathology
Biceps instability test	Shoulder is placed in 90 degrees of abduction and external rotation, then internally rotated while palpating the biceps tendon. A "clunk" or palpable displacement of the tendon is positive test	Subscapularis tear Biceps tendon subluxation due to pulley injury

(continued)

Table 6-1 (continued)

HELPFUL HINTS

Lift-off test	Arm is extended, internally rotated behind the back and the dorsum of the hand placed against the posterior flank. Patient is asked to lift the hand posteriorly off the back. Failure or weakness is positive test	Subscapularis dysfunction or weakness
Belly press test	Palm of the hand on the lower abdomen, with the elbow coplanar. With active compression of the hand on the "belly," the elbow should not move posteriorly	Subscapularis dysfunction or weakness
O'Brien test	Upper extremity is placed in forward flexion to 90 degrees and 15 degrees adduction with the thumb pointed downward. The patient actively resists the examiners downward force. With the arm similarly positioned but the wrist supinated, the patient again resists downward force. Decreased pain in this supinated position is a positive test	Superior labral pathology AC joint pathology
Neer impingement test	Arm passively brought to maximum forward flexion while internally rotated. The test is considered positive if subacromial pain is illicited	Subacromial impingement Rotator cuff tendonitis/tear
Hawkins impingement test	Arm is forward flexed to 90 degrees and the proximal humerus internally rotated	Subacromial impingement Rotator cuff tendonitis/tear

Table 6-2

METHODS FOR EXAMINING THE BICEPS TENDON

Examination	Technique	Illustration	Grading	Significance
Range of motion	Examine in standing and supine positions. Forward flexion, scaption, abduction, internal, and external rotation are documented		Compared bilaterally	Deficiencies may suggest associated pathology such as rotator cuff tears or contractures
Palpation	Patient is standing with the arm supported by the examiner. The proximal humerus is then IR/ER with palpation of the biceps tunnel		Tenderness to palpation within the tunnel is considered a positive test	Helps differentiate between biceps, rotator cuff, AC joint, and muscular tenderness Sensitivity: 53% Specificity: 54%

(continued)

Table 6-2 (continued)

METHODS FOR EXAMINING THE BICEPS TENDON

Examination	Technique	Illustration	Grading	Significance
Speed's test	With the arm in the scapular plane and the wrist supinated, the patient actively forward flexes against resistance		Pain referred to the bicipital groove is positive	Focuses on LHBT Sensitivity: 32% to 90% Specificity: 13.8% to 75%
Yergason's test	With the elbow flexed and the wrist pronated, the patient actively supinates the forearm and flexes the elbow against resistance		Pain referred to the bicipital groove is positive	Focuses on LHBT Sensitivity: 32% Specificity: 75%

(continued)

Table 6-2 (continued)

METHODS FOR EXAMINING THE BICEPS TENDON

Examination	Technique	Illustration	Grading	Significance
Lift-off test	Humerus is internally rotated, the elbow flexed, and dorsum of the hand is on the posterior flank. The patient then attempts to posteriorly separate hand from back		Failure to lift hand off of back is considered a positive lift-off test	Focal assessment of subscapularis Sensitivity: 18% to 92% Specificity: 60% to 98%
Belly press test	The palm is placed on the abdomen and the elbow coplanar. Attempted "belly" press is performed		Posterior drift of the elbow during "belly" press is considered pathologic (positive)	Focal assessment of subscapularis Sensitivity: 40% Specificity: 98%

(continued)

Table 6-2 (continued)

METHODS FOR EXAMINING THE BICEPS TENDON

Examination	Technique	Illustration	Grading	Significance
Active compression test (O'Brien)	Arm is forward flexed to 90 degrees, adducted, and pronated while resisting downward force. The forearm is then supinated, and downward force is again resisted		Decreased pain with resisted down force of the supinated arm is positive for SLAP lesions	Focal test for SLAP tears Sensitivity: 47% to 99% Specificity: 11% to 98%

(continued)

Table 6-2 (continued)

METHODS FOR EXAMINING THE BICEPS TENDON

Examination	Technique	Illustration	Grading	Significance
Neer test	Performed with the extended arm passively brought to maximum forward flexion while internally rotated		Pain referred to the subacromial space is considered positive	Sensitivity: 39% to 89% Specificity: 31% to 98%
Hawkins test	The humerus is forward flexed to 90 degrees, and the humerus is passively internally rotated		Pain referred to the subacromial space is considered positive	Sensitivity: 72% to 95% Specificity: 25% to 66%

Further palpation should be performed of the acromiocla-vicular (AC) and sternoclavicular (SC) joints, clavicle, scapula, rotator cuff insertion, and muscle bellies. The cervical spine is evaluated for focal tenderness, limitations in motion, or neu-rologic symptoms elicited with various positions. The patient is asked to bring the upper extremities through a full, active *range of motion* including forward flexion, scaption, abduction, and internal and external rotation (Figure 6-2). Forward flexion occurs in the sagittal plane and normally ranges from 0 to 150 degrees. Scaption is within the plane of the scapula and requires superior/inferior motion with the arm approximately 20 degrees anterior to the coro-nal plane. Normal values range from 0 to 150 degrees. For external rotation, the elbows are flexed to 90 degrees and are kept at the sides while the humerus rotates. External rotation is also assessed with 90 degrees of shoulder abduction and the forearm initially perpendicular to the coronal plane and slowly rotated to a normal value of approximately 90 degrees. Internal rotation involves the patient reaching for the cra-nial-most aspect of the posterior spine—typically within the mid-thoracic region. Shoulder asymmetry and dysrhythmia are documented. Glenohumeral to scapular motion occurs in an approximate 2:1 ratio with early motion occurring at the glenohumeral joint. Passive range of motion is assessed in both the standing and supine positions.

The Speed's test is considered to provide more focal assess-ment of the biceps tendon. With the elbow flexed 20 degrees and the shoulder forward flexed 60 degrees, a positive test results with resisted forward flexion and pain referred to the bicipital groove. This test remains a popular clinical tool for evaluating the LHBT; however, concerns remain as sensitivity ranges from 32% to 90%, with specificity ranges of 13.8% to 75%.[8,9]

The Yergason's test is another physical exam maneuver directed at the LHBT. With the elbow flexed to 90 degrees and the forearm pronated, the patient is asked to actively flex the elbow and supinate the wrist against resistance. Pain referred to the bicipital groove is considered a positive test. Sensitivity for this test is 32% with a specificity of 75%.[9]

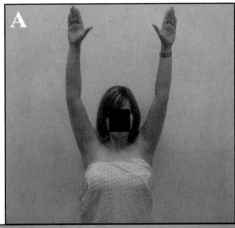

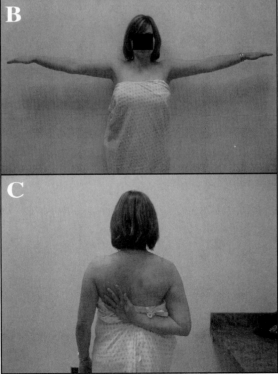

Figure 6-2. Range of motion. Bilateral passive and active range of motion is measured. (A) Forward flexion. (B) Abduction. (C) Internal rotation (Apley scratch test) *(continued).*

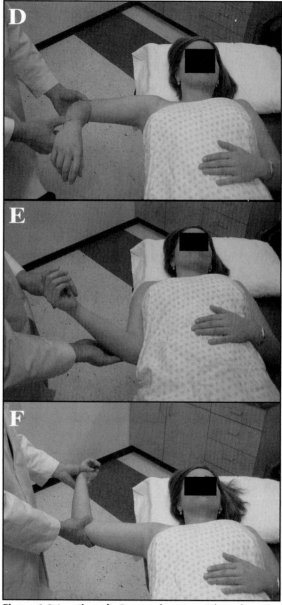

Figure 6-2 (continued). Range of motion. Bilateral passive and active range of motion is measured. (D) Internal rotation at 90 degrees abduction. (E) External rotation demonstrated at neutral (0 degrees abduction). (F) External rotation demonstrated at 90 degrees of abduction.

The biceps instability test represents a dynamic maneuver that assesses the stability of the LHBT within the biceps tunnel (Figure 6-3). The patient's shoulder is placed in 90 degrees of abduction and external rotation and then internally rotated while palpating the biceps tendon. A "clunk" or palpable displacement of the tendon is considered positive for dysfunction of its tunnel restraints—most likely a subscapularis tear.[10]

Given the association between rotator cuff tears and dysfunction of the biceps tendon, a comprehensive exam of the rotator cuff is performed. Shoulder girdle and rotator cuff *motor strength* are tested bilaterally. Grading is based on the 5-point Medical Research Council (MRC) Motor Strength Scale: 5/5 = active movement against full resistance; 4/5 = active movement with weakness to resistance; 3/5 = active movement to gravity only; 2/5 = active movement without gravity; 1/5 = trace contraction; 0/5 = no contraction.[11] Particular attention is paid toward the subscapularis as tears in this area are most commonly associated with biceps subluxation. Several tests allow for the assessment of subscapularis function, including the lift-off test and belly press test.

Originally described in 1991, the lift-off test allows for focal evaluation of subscapularis strength.[10] With the arm extended and internally rotated and the dorsum of the hand placed against the posterior flank, the patient is asked to lift the hand posteriorly off the back. Failure or weakness suggests a tear or dysfunction of the subscapularis. Sensitivity of the test ranges from 18% to 92% and specificity ranges from 60% to 98%.[10,12,13]

The *belly press test* provides another technique for evaluation of the subscapularis.[14] The patient places the palm of the hand on the lower abdomen with the elbow coplanar. With active compression of the hand on the "belly," the elbow should not move posteriorly. A positive test results as the subscapularis cannot maintain internal rotation of the humerus and the elbow drifts backward. Sensitivity of this test is 40% and specificity is 98%.[12]

Given the course of the biceps tendon and its insertion on the superior labrum, it is necessary to assess the patient for SLAP lesions. The *O'Brien test* involves comparison between 2 positions of resistance. The patient's upper extremity is placed in forward flexion to 90 degrees and adduction with

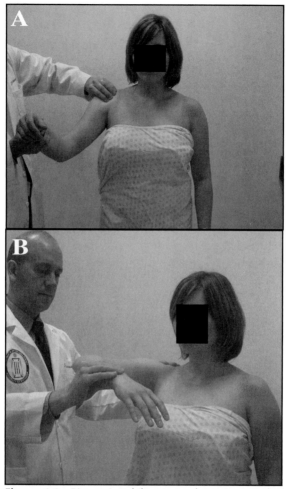

Figure 6-3. Biceps instability test—dynamic test that assesses for biceps subluxation or dislocation tendon from the bicipital groove. (A) The biceps tendon is identified within the bicipital groove with the arm abducted and externally rotated. (B) The arm is then internally rotated with continued palpation of the bicep.

the thumb pointed inferior. The patient actively resists the examiner's downward force. With the arm similarly positioned but the wrist supinated, the patient again resists downward force. Decreased pain in this supinated position suggests a SLAP tear. Sensitivity of the maneuver ranges from 47% to 99%, with a specificity of 11% to 98%.[15-17]

Impingement testing is usually assessed with *Neer test* and *Hawkins test*.[18,19] Anatomic studies have shown that with these maneuvers, multiple structures come into contact with the coracoacromial arch, including the biceps tendon, rotator cuff, and greater and lesser tuberosities of the humerus.[20] Furthermore, these tests appose the undersurface of the rotator cuff with the glenoid rim, making isolation of specific pathology even more difficult. *Neer sign* is performed with the arm passively brought to maximum forward flexion while internally rotated. The test is considered positive if subacromial pain is elicited. Sensitivity for this test ranges from 39% to 89%, with a specificity of 31% to 98%.[21,22] *Hawkins sign* was originally described as being a "less reliable" method of impingement assessment.[18] Meta-analysis actually shows the test to be more reliable than *Neer sign* with a sensitivity range of 72% to 95% and a specificity of 25% to 66%.[22-24] The test is performed with the arm forward flexed to 90 degrees and the proximal humerus internally rotated.

PATHOANATOMY

Dysfunction of the LHBT may be categorized as inflammatory, traumatic, or instability related. Given the confines of the biceps tunnel, restraint to humeral head migration and translation, and the stresses encountered at the insertion onto the glenoid, it is not surprising that inflammatory conditions contribute to the majority of LHBT pathology.

Originating on the supraglenoid tubercle and confluent with the superior labrum, the LHBT traverses the glenohumeral joint and enters the bicipital groove. Within the bicipital groove, the proximal part of the LHBT is stabilized by the reflection pulley composed of 4 soft tissues—superior glenohumeral ligament (SGHL), coracohumeral ligament (CHL), subscapularis tendon, and supraspinatus tendon. The roof of

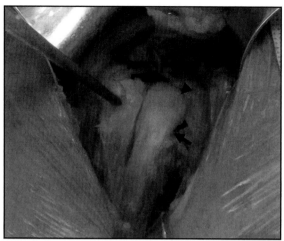

Figure 6-4. Diseased proximal biceps tendon. The LHBT is identified within its sheath. Note the hypertrophic proximal portion (arrowhead) and the constricted inter-tubercular portion of the tendon (arrow).

the bicipital groove is composed of the CHL and transverse humeral ligament. The floor is composed of the SGHL. The subscapularis and supraspinatus augment the reflection pulley with tendinous contributions.

Even though the LHBT passes through the glenohumeral joint, it is considered to be extrasynovial due to its envelope. Inflammatory conditions affecting the synovial lining potentiate pain as it extends into the constrained biceps tunnel with limited room to accommodate swelling (Figure 6-4). Distal to this tunnel, its muscular contribution to the biceps coalesces with that of the other proximal head—the short head.

Trauma-related biceps pathology may be related to a single event but is more commonly associated with repetitive microtrauma. Overhead athletes, such as baseball pitchers and swimmers or laborers, typically comprise this population as there is a particular tension load to the biceps tendon during arm deceleration. Although SLAP lesions represent the majority of longitudinal traction injuries, single traumatic events may occur within the substance of the LHBT and lead to a tear.

In situations involving instability of the LHBT, abnormalities related to the constraints, or pulley system, of the biceps tunnel lead to the pathologic subluxation or dislocation of the tendon. Proximal humerus fractures that disrupt the bony constraints of the intertubercular groove may lead to instability. The most common direction for dislocation is anteroinferior with loss of the underlying SGHL. As the LHBT subluxes out of the reflection pulley and tunnel, added stress on the superior edge of the subscapularis may lead to tears within this rotator cuff tendon.

IMAGING

Radiographs remain a standard component during the initial evaluation of patients with shoulder pain. Routine films include a true AP perpendicular to the scapular plane, an axillary view, and an outlet view. Although rarely obtained, the Fisk view offers axial imaging of the proximal humerus, displaying the borders of the bicipital groove. Additional views may be obtained as dictated by the clinical exam. Plain films provide limited information related to pathology of the biceps tendon and have a greater role in identifying or ruling out concomitant shoulder issues.

Magnetic resonance imaging (MRI) is the most valuable imaging modality for evaluating the biceps tendon. The course and constraints of the tendon may be appreciated in multiple planes and deficiencies recognized. Additional pathologies are readily identified and allow for a more comprehensive therapeutic approach (Figure 6-5). The addition of contrast (ie, MRI arthrography) allows for further delineation on shoulder pathology.[25]

TREATMENT

Conservative

Therapeutic options for the treatment of proximal biceps-related dysfunction range from conservative to surgical. Rarely is the biceps an isolated pathology, and it is necessary

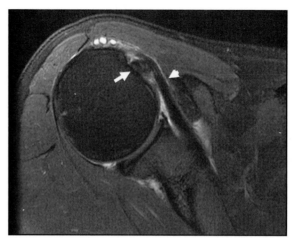

Figure 6-5. MRI (axial PD FSE) illustrating tendonitis of the LHBT (arrow) with subluxation under the subscapularis tendon (arrowhead).

to address the concomitant shoulder issues to maximize outcome. For primary bicipital tendonitis, a nonsurgical approach is initially pursued. A trial of rest, anti-inflammatory medications, and activity modification is recommended. Injections into the biceps sheath or tunnel may provide relief; however, there is risk of iatrogenic rupture if the tendon itself is violated.[26] Patients may also benefit from subacromial injections. For focused rehabilitation, it is critical to delineate the contributors to the shoulder pain given that 95% of biceps tendonitis is secondary.[2] In the situation of biceps instability, rehabilitation will be of little benefit, and symptomatic treatment remains the best, nonsurgical option. Failure of conservative modalities after 3 months is an indication for more aggressive intervention.[27]

Surgical

Depending on the degree of biceps damage and associated conditions, surgical modalities include débridement, tenotomy, tenodesis, and decompression. Accepted criteria for surgical management are related to what are considered irreversible changes in the biceps tendon: >25% fraying, subluxation from the bicipital groove, or a decrease in tendon size by

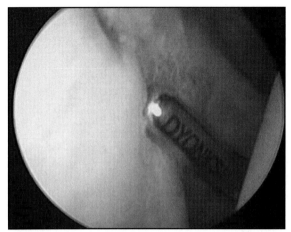

Figure 6-6. Arthroscopic biceps débridement. View from posterior portal.

25%.[28] These are typically performed in settings of additional shoulder reconstruction procedures including rotator cuff and labral repair, as well as subacromial decompression. Although the majority of surgical interventions involve arthroscopy, open procedures may be performed depending on surgeon comfort.

Biceps débridement is a reasonable intervention for biceps tears involving <50% of the intra-articular tendon, particularly in sedentary individuals. Synovitis of the tendinous sheath must also be limited, as extensive inflammation suggests an evolving process (Figure 6-6). For younger, more active individuals, débridement creates a weaker construct, leaving the remaining tendon more susceptible to reaggravation and additional procedures. Within this population, only fraying of <25% should be pursued with débridement. This is most easily performed arthroscopically, as visualization is optimum and the recovery fast.

A popular surgical approach involves a tenotomy, or release, of the biceps tendon from its origin on the superior labrum or glenoid (Figure 6-7). Particularly attractive is the ease with which it can be performed arthroscopically. Candidates generally include older individuals with extensive biceps-related pain who are comfortable with the potential

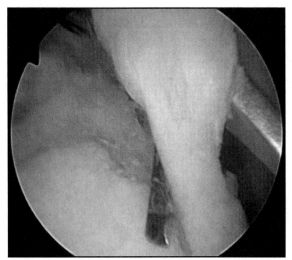

Figure 6-7. Arthroscopic biceps tenotomy. View from posterior portal.

cosmetic "Popeye" deformity. Postoperative complaints related to muscle cramping and fatigue are common in younger adults, and patients should be well informed prior to surgical release. This discomfort is uncommon in individuals older than 60 years.[29] Most studies looking at the outcomes of tenotomy are in the setting of palliative management of large rotator cuff tears with subjective good to excellent results in 86% of patients.[30] Contraindications to tenotomy, as well as tenodesis, include patients with true pseudoparalysis of the shoulder with rotator cuff tears.

Tenodesis options are attractive because they attempt to preserve anatomy: tension, length, cosmesis, and minimal loss of elbow flexion and supination strength while removing the diseased portion of the tendon (Figure 6-8). Disadvantages are the higher complexity of the procedure and prolonged immobility and rehabilitation. Candidates include all patients with significantly (>25%) diseased biceps tendons, but it is especially recommended for younger individuals who wish to remain active. Multiple techniques have been described and range from soft tissue tenodesis to procedures that anchor the

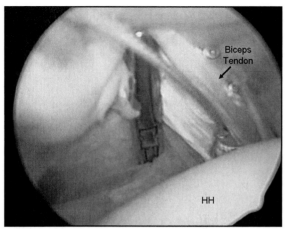

Figure 6-8. Arthroscopic biceps tenodesis. View from posterior portal. Suture placement into the biceps tendon (labeled) with tenodesis to the transverse humeral ligament. HH = humeral head.

released tendon to bone. The tenodesis may be proximally or distally based.

On rare occasions, a biceps decompression may be an appropriate intervention when constriction of the intra-articular synovial sheath or transverse humeral ligament covering the bicipital tunnel becomes pathologic. When the tunnel is involved, open procedures are required for the surgical approach; otherwise, it may be addressed arthroscopically.

CONCLUSION

Disorders of the LHBT contribute significantly to shoulder pain. The diagnosis remains difficult and must be approached with a detailed understanding of the anatomy, an appreciation of the associated soft tissue and bony pathology, and a comfort with various physical exam techniques in order to properly identify, treat, and optimize the outcome of patients with dysfunction of the LHBT.

REFERENCES

1. Curtis AS, Snyder SJ. Evaluation and treatment of biceps tendon pathology. *Orthop Clin North Am.* 1993;24:33-43.
2. Post M, Benca P. Primary tendinitis of the long head of the biceps. *Clin Orthop Relat Res.* 1989;(246):117-125.
3. Furlani J. Electromyographic study of the m. biceps brachii in movements at the glenohumeral joint. *Acta Anat (Basel).* 1976;96:270-284.
4. Warner JJP, McMahon PJ. The role of the long head of the biceps brachii in superior stability of the glenohumeral joint. *J Bone Joint Surg Am.* 1995;77:336-372.
5. Itoi E, Kuechle DK, Newman SR, Morrey BF, An KN. Stabilising function of the biceps in stable and unstable shoulders. *J Bone Joint Surg Br.* 1993;75(4):546-550.
6. Rodosky MW, Harner CD, Fu FH. The role of the long head of the biceps muscle and superior glenoid labrum in anterior stability of the shoulder. *Am J Sports Med.* 1994;22(1):121-130.
7. Gill HS, El Rassi G, Bahk MS, Castillo RC, McFarland EG. Physical examination for partial tears of the biceps tendon. *Am J Sports Med.* 2007;35(8):1334-1340.
8. Bennett WF. Specificity of the Speed's test: arthroscopic technique for evaluating the biceps tendon at the level of the bicipital groove. *Arthroscopy.* 1998;8:789-796.
9. Holtby R, Razmjou H. Accuracy of the Speed's and Yergason's tests in detecting biceps pathology and SLAP lesions: comparison with arthroscopic findings. *Arthroscopy.* 2004;20(3):231-236.
10. Gerber C, Krushell RJ. Isolated rupture of the tendon of the subscapularis muscle. Clinical features in 16 cases. *J Bone Joint Surg Br.* 1991;73:389-394.
11. Medical Research Council. *Aids to Examination of the Peripheral Nervous System. Memorandum no. 45.* London, UK: Her Majesty's Stationary Office; 1976.
12. Barth JR, Burkhart SS, De Beer JF. The bear-hug test: a new and sensitive test for diagnosing a subscapularis tear. *Arthroscopy.* 2006;22:1076-1084.
13. Hertel R, Ballmer FT, Lombert SM, et al. Lag signs in the diagnosis of rotator cuff rupture. *J Shoulder Elbow Surg.* 1996;5:307-313.
14. Gerber C, Hersche O, Farron A. Isolated rupture of the subscapularis tendon. *J Bone Joint Surg.* 1996;78A:1015-1023.
15. McFarland EG, Kim TK, Savino RM. Clinical assessment of three common tests for superior labral anterior-posterior lesions. *Am J Sports Med.* 2002;30:810-815.
16. Myers TH, Zemanovic JR, Andrews JR. The resisted supination external rotation test: a new test for the diagnosis of superior labral anterior posterior lesions. *Am J Sports Med.* 2005;33:1315-1320.
17. O'Brien SJ, Pagnani MJ, Fealy S, et al. The active compression test: a new and effective test for diagnosing labral tears and acromioclavicular joint abnormality. *Am J Sports Med.* 1998;26:610-613.

18. Hawkins RJ, Kennedy JC. Impingement syndrome in athletes. *Am J Sports Med.* 1980;8:151-157.
19. Neer CS II. Anterior acromioplasty for the chronic impingement syndrome in the shoulder. *J Bone Joint Surg.* 1972;54A:41-50.
20. Valadie AL III, Jobe CM, Pink MM, et al. Anatomy of provocative tests for impingement syndrome of the shoulder. *J Shoulder Elbow Surg.* 2000;9:36-46.
21. Bak K, Faunl P. Clinical findings in competitive swimmers with shoulder pain. *Am J Sports Med.* 1997;25:254-260.
22. Calis M, Akgun K, Birtane M, et al. Diagnostic values of clinical diagnostic tests in subacromial impingement syndrome. *Ann Rheum Dis.* 2000;59:44-47.
23. Hegedus EJ, Goode A, Campbell S, et al. Physical examination tests of the shoulder: a systematic review with meta-analysis of individual tests. *Br J Sports Med.* 2008;42(2):80-92.
24. Park HB, Yokota A, Gill HS, et al. Diagnostic accuracy of clinical tests for the different degrees of subacromial impingement syndrome. *J Bone Joint Surg Am.* 2005;87:1446-1455.
25. Zanetti M, Weishaupt D, Gerber C, Hodler J. Tendinopathy and rupture of the tendon of the long head of the biceps brachii muscle: evaluation with MR arthrography. *AJR Am J Roentgenol.* 1998;170(6):1557-1561.
26. Burkhead WZ Jr, Arcand MA, Zeman C, Habermeyer P, Walch G. The biceps tendon. In: Rockwood CA Jr, Matsen FA III, Wirth MA, Lippitt SB, eds. *The Shoulder.* Vol 2. 3rd ed. Philadelphia, PA: Saunders; 2004:1059-1119.
27. Barber A, Field LD, Ryu R. Biceps tendon and superior labrum injuries: decision marking. *J Bone Joint Surg Am.* 2007;89(8):1844-1855.
28. Sethi N, Wright R, Yamaguchi K. Disorders of the long head of the biceps tendon. *J Shoulder Elbow Surg.* 1999;8:644-654.
29. Gill TJ, McIrvin E, Mair SD, et al. Results of biceps tenotomy for treatment of pathology of the long head of the biceps brachii. *J Shoulder Elbow Surg.* 2001;10:247-249.
30. Walch G, Edwards TB, Boulahia A, et al. Arthroscopic tenotomy of the long head of the biceps in the treatment of rotator cuff tears: clinical and radiographic results of 307 cases. *J Shoulder Elbow Surg.* 2005;14:238-246.

7

ACROMIOCLAVICULAR JOINT

Michael Shin, MD and Geoffrey S. Baer, MD, PhD

INTRODUCTION

The acromioclavicular (AC) joint is a common source of shoulder pain and should be evaluated as a part of a routine shoulder exam. Whether due to idiopathic osteoarthritis in a middle-age patient, to a traumatic event in a football player, or to insidious-onset shoulder pain in a heavy weightlifter, the evaluation of the AC joint plays a prominent role in the proper diagnosis and treatment of patients with shoulder pain (Table 7-1).

Approximately 9% of injuries to the shoulder girdle involve the AC joint, and the common conditions that afflict the AC joint include osteoarthritis, post-traumatic arthritis,

Cohen SB. *Musculoskeletal Examination of the Shoulder:
Making the Complex Simple* (pp. 136-160).
© 2011 SLACK Incorporated

Table 7-1

HELPFUL HINTS

AC Joint Disorders

1. Osteoarthritis: Idiopathic. Arthrosis is common but frequently asymptomatic. Pain localized to AC, tender to palpation, pain with cross-body adduction. Selective lidocaine injection into AC joint helpful in differentiating AC symptoms from other sources of shoulder pain and is also a strong predictor of response to surgical treatment. Nonoperative treatment is usually successful. For patients who fail nonoperative treatment, distal clavicle excision has reliable and excellent results.

2. Post-traumatic arthritis: History of shoulder trauma, most commonly intra-articular clavicle fractures or type I and II AC separations. Symptoms, physical exam, and treatment are similar to idiopathic osteoarthritis.

3. Distal clavicle osteolysis: Repetitive microtrauma commonly from weightlifting. Exercises that frequently aggravate symptoms include bench press, dips, push-ups, flies, and overhead lifting. Physical exam findings and treatment are similar to osteoarthritis.

4. AC separations: Most common mechanism is a fall directly onto the acromion with the humerus adducted. Frequently found in contact athletes (football, hockey, rugby). Type is based on magnitude and direction of displacement, and treatment is based on type. Types I and II are predominately treated nonoperatively. Types IV, V, and VI are treated surgically. Treatment of Type III separations is controversial but usually nonoperative.

Imaging

1. X-rays
   a. Shoulder series (AP, scapular Y, axillary): To rule out concomitant shoulder pathology, including dislocation, glenohumeral arthritis, and fractures. Axillary view can help visualize anteroposterior translation in separations.
   b. Zanca view: Ten to 15 degree cephalic tilt and 50% normal penetrance. Best view for AC joint pathology. Obtain both AC joints on single film to evaluate displacement in AC separations.
   c. Stress views: Manual traction or weights to dependent arm to help delineate AC separations. Rarely used due to limited utility and patient discomfort.
   d. Bone scan: Due to high rate of asymptomatic AC arthrosis, can help identify symptomatic AC degeneration.
   e. MRI: Allows evaluation of soft tissues, provides improved anatomic detail, and enables evaluation of other common areas of shoulder pathology including the labrum, rotator cuff, and biceps tendon.

separations, and distal clavicular osteolysis. The AC joint is commonly injured in competitive and recreational athletes given its superficial and vulnerable location, especially in athletes who participate in contact sports such as football, hockey, and rugby. Separation of the AC joint occurs most often in the first 3 decades of life and is more common in men by a 5:1 ratio. AC separations are also more likely to be incomplete than complete by a 2:1 ratio.[1] Shoulder injuries accounted for 15% of injuries among Division I college hockey athletes, and in these players, AC joint separations were the most common shoulder injury and third most common injury overall.[2] In elite college football players, shoulder injuries comprised the fourth most common musculoskeletal injury, and AC separations were the most common of these shoulder injuries, accounting for 41%.[3] For NFL quarterbacks, shoulder injuries were found to be the second most common injury at 15.2%, just behind head injuries, and AC separations were the most common injury, accounting for 40% of all shoulder injuries.[4] AC separations are also common in recreational alpine skiers, composing 19.6% of all shoulder injuries.[5]

Due to the relatively large forces transmitted across its small surface area, the AC joint is commonly involved in degenerative conditions, though frequently asymptomatic. The hyaline cartilage and intrameniscal disk of the AC joint can deteriorate due to mechanical wear from repetitive motion or articular incongruity from prior trauma and lead to AC joint arthrosis. Less commonly, distal clavicular osteolysis can cause subchondral cystic changes, degeneration, and relative osteopenia as seen in patients with repetitive microtrauma to the AC joint, such as in weightlifters.[6,7]

Most importantly, shoulder pain is often multifactorial, and AC joint pathology can occur both in isolation and in concert with other common shoulder maladies. The evaluation of the shoulder is further complicated by the fact that though AC joint degeneration and pathology is common, it is also commonly asymptomatic. The evaluation of the AC joint through history, physical examination, and imaging will be reviewed to properly diagnose and define the AC joint's role in a patient's shoulder pain and dysfunction. Then, treatment of the AC joint, both nonoperative and operative, including open and arthroscopic treatments as well as early versus delayed surgery, will be discussed.

HISTORY

AC joint pathology most often presents with pain in the superior and/or anterior shoulder. Activities that involve reaching across the body, reaching overhead, and pushing off can be particularly symptomatic. Most AC osteoarthritis is idiopathic with no distinct traumatic mechanism, and patients will typically describe dull or aching pain that is insidious in onset.[6,7]

Post-traumatic AC joint arthritis is more common than idiopathic osteoarthritis, and patients may have a history of a shoulder injury but it is typically seen in patients with a history of intra-articular distal clavicle fractures or type I and II AC separations. Some natural history studies of type I and II AC separations have shown symptomatic post-traumatic arthritis rates as high as 42%.[8,9]

The most common mechanism of injury for AC separations is a fall directly onto the acromion with the humerus in an adducted position. The force applied directly onto the acromion displaces the acromion inferomedially, causing disruption of the AC ligaments and capsule. The severity of the separation increases as greater forces are applied and can further disrupt the coracoclavicular (CC) ligaments, the deltoid and trapezius muscle attachments, and the deltotrapezial fascia. A less common mechanism of injury involving an indirect trauma has also been described in which a fall on an outstretched hand or elbow transmits force through the humerus and up to the acromion.[10]

The rate of osteoarthritis of the AC joint increases with increasing age, and reported rates of radiographic evidence of AC osteoarthrosis are as high as 57% in the elderly population, most of which are asymptomatic.[11] Traumatic afflictions of the AC joint, however, are more frequent in the younger population. Distal clavicular osteolysis is also more common in younger, male populations and is most common in weight-lifters. Osteolysis can affect as many as 28% of elite weightlifters, but can be seen in any patient or athlete who is involved in repetitive activity or heavy lifting. The pain is insidious in onset and localizes to the anterior and/or superior shoulder with swelling and AC tenderness. These patients may complain of more discomfort during certain activities, such as push-ups, bench-press, flies, overhead lifting, and dips. In several studies, at the time of presentation, up to 79% of weight-lifters were found to have bilateral AC joint osteolysis.[12,13]

EXAMINATION

The examination of the shoulder involves a thorough and systematic evaluation of not just the glenohumeral and AC joints but the cervical spine as well (Table 7-2). Many shoulder pathologies are found concomitantly, and all sources of shoulder dysfunction and pain should be delineated in a patient complaining of shoulder pain. Patients with shoulder pain with possible AC joint pathology should also have thorough evaluations for instability (see Chapter 3), labral tears (see Chapter 4), rotator cuff tears (see Chapter 5), impingement, and biceps tendon disorders (see Chapter 6).

As part of a complete shoulder examination, several tests are commonly used to evaluate for AC joint pathology. Inspection of the AC joint may reveal prominence and/or deformity due to swelling or osteophytes. In the setting of an AC separation, patients may present with swelling, deformity, and ecchymosis. The deformity may be very significant in patients with higher type separations. Evaluation of range of motion (ROM) is usually normal but may reveal crepitus and pain localizing to the AC joint, especially with horizontal adduction and overhead motion. Palpation typically elicits tenderness of the AC joint and possibly distal clavicle and is the most sensitive test for AC joint pathology as demonstrated in multiple studies.[14-16] The cross-arm adduction test is also a reliable test specifically for AC joint pathology and is performed both actively and passively. The test loads the joint and reproduces pain that localizes to the AC joint.[14-16] Pain can also localize to the AC joint during impingement testing (Neer and Hawkins tests); however, because pain is frequently present in patients with other common shoulder pathologies, such as impingement, rotator cuff tears, and superior labral tears, positive impingement tests alone are less specific when evaluating patients for potential AC joint pathology. The O'Brien active compression test was originally described as effective for diagnosing both labral tears and AC joint pathologies.[17]

Table 7-2

METHODS FOR EXAMINING THE ACROMIOCLAVICULAR JOINT

Examination of the Shoulder for Acromioclavicular Joint Pathology

EXAM	TECHNIQUE	ILLUSTRATION	SIGNIFICANCE
Inspection	Examine for AC prominence, swelling, ecchymosis, and deformity		Swelling and prominence common with degenerative change. Prominence, swelling, ecchymosis, and deformity may be due to AC separation
Palpation	Tenderness with direct palpation of the AC joint. Reducability of AC joint		One of the most sensitive tests for AC joint pathology

(continued)

Table 7-2 (continued)

METHODS FOR EXAMINING THE ACROMIOCLAVICULAR JOINT

EXAM	TECHNIQUE	ILLUSTRATION	SIGNIFICANCE
Range of motion	Test active and passive ROM in all planes and compare to contralateral side	  	ROM is usually maintained but often painful at the AC joint when above shoulder height
Cross-arm adduction	Shoulder flexed to 90 degrees and adducted across the body—both active and passive		Pain localizing to AC joint is a reliable test for AC joint pathology

(continued)

Table 7-2 (continued)

Methods for Examining the Acromioclavicular Joint

Exam	Technique	Illustration	Significance
O'Brien active compression	Shoulder flexed to 90 degrees with 30 degrees of adduction and in full internal rotation resisting examiner's downward force. The arm is then fully externally rotated, and the patient again resists downward force		Anterosuperior pain with arm internally rotated that is improved with external rotation is positive for AC joint pathology
AC joint compression (Paxinos test)	Examiner uses thumb and index or long finger to compress the posterolateral acromion toward the midpart of the clavicular shaft		Pain or increased pain in the AC joint is positive for pathology
Differential lidocaine injection	Injection of local anesthetic (± corticosteroid) into AC joint		Pain relief is highly accurate for AC joint pathology. Helps differentiate AC joint pain from other sources of pain. Most accurate predictor of successful outcome from surgical management.

The test is performed with the patient's arm in 90 degres of forward flexion, 30 degrees of adduction, and in full internal rotation. The patient then resists a downward force placed by the examiner on the patient's forearm. With the arm maintained in the same position, the patient now resists the examiner's force with the shoulder fully externally rotated. A positive test for AC joint pathology is when the first maneuver causes superficial or anterior shoulder pain that is less intense or alleviated with the second maneuver.[17] The AC joint compression test (the Paxinos test) evaluates the AC joint and is positive for AC joint pathology when the patient experiences pain or increased pain as the examiner compresses the posterolateral aspect of the acromion toward the midpart of the clavicular shaft between the thumb and index or long finger.[16] Differential lidocaine injection tests, with or without corticosteroid, can be both diagnostic and therapeutic. They can also help differentiate multiple sources of shoulder pain and can increase diagnostic accuracy in equivocal cases. Symptomatic relief from a lidocaine injection into the AC joint is the most accurate predictor of successful outcome from surgical management, and an AC joint injection's failure to relieve or improve a patient's symptoms should raise suspicion to the diagnosis and to the potential effectiveness of operative treatment.[18,19]

PATHOANATOMY

The AC joint is a diarthrodial joint with a fibrocartilaginous meniscal disk connecting the medial facet of the acromion and the distal clavicle. The AC joint capsule is reinforced and stabilized by anterior, posterior, superior, and inferior ligaments, of which the superior AC ligament is the thickest and strongest. The superior AC ligament is further supported by fibers from the attachments of the deltoid and trapezius muscles.[20] The intra-articular meniscal disk has a wide variation in size and shape and degenerates rapidly with age. The meniscal disk is thought to have a minimal function in the joint.[21]

AC ligaments, especially the superior and posterior ligaments, are the primary restraint to anteroposterior translation and also contribute to superior stability. The CC ligaments, the conoid and trapezoid ligaments, provide the primary restraint to superior translation of the AC joint with the conoid ligament functioning as the major contributor. The trapezoid ligament origin is 26 mm medial to the AC joint on the mid-portion of the undersurface of the clavicle, while the conoid ligament origin is more posterior and is 46 mm medial to the AC joint.[22]

The AC joint can experience large forces as one of the few attachments connecting the appendicular to the axial skeleton at the shoulder girdle. This transmission of force contributes to the high prevalence of osteoarthrosis with increasing age, which does not necessarily correlate with symptoms. Post-traumatic arthritis may develop in patients with prior clavicle fractures, especially intra-articular fractures and in patients with AC joint separations, particularly type I and II separations.[23] Osteolysis is a relatively uncommon cause of AC joint pain and is most commonly due to repetitive microtrauma, such as seen in weightlifters, that results in pain and resorption of the distal clavicle. AC separations are most commonly sustained with a fall directly onto the top of the acromion with the humerus in an adducted position. AC separations are typically classified using the Rockwood modification of the Tossy and Allman Classification based upon the magnitude and direction of clavicular displacement[1] (Table 7-3).

IMAGING

When evaluating a patient with shoulder pain, a standard shoulder series with AP, scapular Y, and axillary views are recommended (see Table 7-1). To specifically evaluate the AC joint, the Zanca view best visualizes AC joint pathology. This view is obtained with a 10- to 15-degree cephalic tilt and 50% normal penetrance.[24] Osteoarthritis and post-traumatic arthritis will demonstrate the typical radiographic findings of arthrosis, including joint space narrowing, osteophyte formation, subchondral cysts, and sclerosis (Figure 7-1). Radiographs of distal clavicular osteolysis will reveal resorption with widening of the joint space, subchondral cystic changes, and relative osteopenia (Figure 7-2).

Table 7-3

ROCKWOOD MODIFICATION OF *TOSSY* AND *ALLMAN* CLASSIFICATION OF *ACROMIOCLAVICULAR SEPARATIONS*

Acromioclavicular Separation Types

TYPE	X-RAY FINDINGS	STRUCTURAL DISRUPTION	ILLUSTRATION
I	Normal AC joint width and CC inter-space	Sprain of AC ligaments but functionally competent	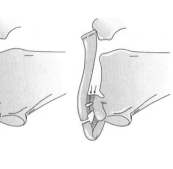
II	<100% superior displacement of dis-tal clavicle	AC ligaments torn, CC ligaments sprained	

(continued)

Table 7-3 (continued)

Rockwood Modification of Tossy and Allman Classification of Acromioclavicular Separations

Type	X-ray Findings	Structural Disruption	Illustration
III	100% displacement of distal clavicle	AC ligaments torn, CC ligaments disrupted	
IV	CC interspace may be increased on AP view, axillary view reveals posterior translation of clavicle	AC and CC ligaments disrupted, distal clavicle displaced posteriorly into or through trapezius muscle	

(continued)

Table 7-3 (continued)

ROCKWOOD MODIFICATION OF TOSSY AND ALLMAN CLASSIFICATION OF ACROMIOCLAVICULAR SEPARATIONS

Type	X-ray Findings	Structural Disruption	Illustration
V	100% to 300% superior displacement of distal clavicle, with CC interspace increased 2 to 3 times	AC and CC ligaments disrupted, distal clavicle buttonholed through deltotrapezial fascia	
VI	Dislocation of distal clavicle into a subacromial or subcoracoid position, CC interspace is reduced	AC ligaments are torn, CC ligaments and trapezius muscle attachment disruptions are variable	

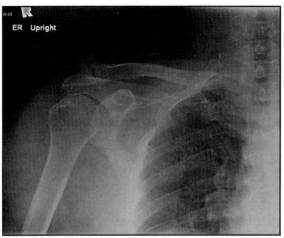

Figure 7-1. Zanca view of right shoulder AC joint degenerative joint disease.

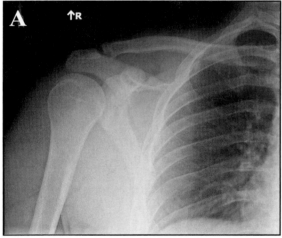

Figure 7-2. Distal clavicle osteolysis. (A) Radiograph showing cysts and osteopenia of the distal clavicle *(continued)*.

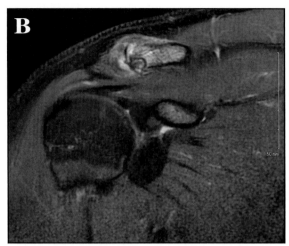

Figure 7-2 (continued). Distal clavicle osteolysis. (B) Coronal MRI showing cystic changes, swelling, bone edema, and capsular hypertrophy.

Radiographs for AC separations are best obtained with an AP view with both AC joints visualized on a single film. An axillary view can also be obtained to assess AP translation of the AC joint. Stress radiographs have been used to confirm instability and to differentiate between type II and III AC separations. Stress radiographs are obtained by applying inferior traction to the dependent arms, either with weights or manual traction. Stress views are now rarely used, as the usefulness of this view in differentiating types has come into question and the information garnered is not thought to outweigh the additional cost, patient discomfort, and time consumption.[25]

Given the high rate of asymptomatic AC arthrosis, bone scan has been advocated by some authors as a modality of choice because it is more sensitive in evaluating symptomatic degeneration of joints, including the AC joint.[16]

MRI has been proposed in evaluating AC injuries, especially AC separations due to the ability to evaluate the soft tissues and the higher level of anatomic detail. MRI findings of AC joint arthrosis evaluate the joint for osteophytes, subchondral cysts and irregularities, bone marrow edema, joint effusion, and joint capsule hypertrophy (Figure 7-3).[26,27]

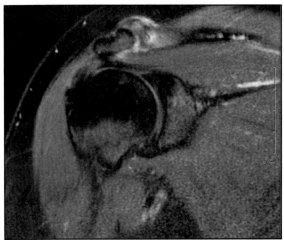

Figure 7-3. Coronal MRI of AC degenerative joint disease.

TREATMENT

Treatment options for osteoarthritis, post-traumatic arthritis, and distal clavicular osteolysis start with nonoperative management. Nonsteroidal anti-inflammatory drugs, physical therapy to strengthen the rotator cuff and to restore normal range of motion, activity modification, and local anesthetic/corticosteroid injections are the cornerstones of nonoperative treatment. Activity modification ranges from complete avoidance of the aggravating or inciting activities to relative rest or modification of exercise technique to relieve symptoms.

For patients who have undergone 3 to 6 months of nonoperative treatment and continue to be symptomatic or are unable or unwilling to modify inciting activities, operative treatment involving resection of the distal clavicle is considered. Patients who obtain relief from a lidocaine injection into the AC joint have the best prognosis for relief from surgical management. Operative treatment accomplished by open or arthroscopic means has been shown to have similar results. Open resection was first reported by Mumford in 1941 and independently in the same year by Gurd and is advocated for generally shorter operating room times, being less technically demanding,

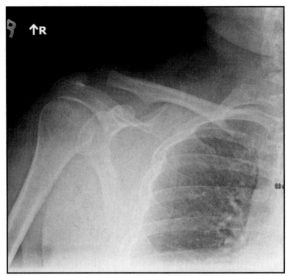

Figure 7-4. Postoperative radiograph of an open distal clavicle resection.

and better visualization to ensure adequate resection (Figure 7-4).[28,29] Arthroscopic distal clavicle resection was reported by Johnson in 1986 and has been advocated due to improved cosmesis, avoiding muscular detachment, decreased pain, and earlier rehabilitation (Figure 7-5).[30]

Different surgeons and researchers have recommended distal clavicle resections ranging from as little as 4 mm to greater than 2.5 cm. It is currently recommended and accepted that regardless of technique, open or arthroscopic, 7 to 10 mm of distal clavicle be resected. Care must be taken with all techniques to avoid over-resection of the distal clavicle including the CC ligaments or superior AC ligaments because this could result in postoperative instability. However, incomplete or inadequate resection of the distal clavicle should also be avoided, which would cause recurrence of symptoms. Extra scrutiny must be taken when performing arthroscopic resections as visualization may be more difficult than with the open technique. Frequently, inadequate resection with the arthroscopic technique is at the posterior aspect of the distal clavicle and can lead to posterior abutment and residual pain.

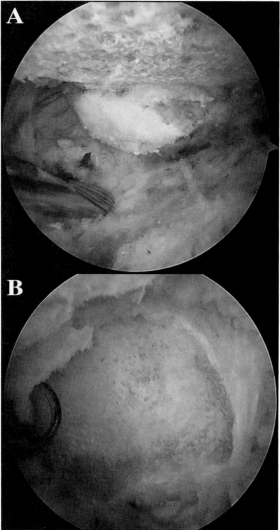

Figure 7-5. Arthroscopic distal clavicle resection. (A) Arthritic distal clavicle prior to excision. Viewed from posterior portal. (B) Arthroscopic resection of the distal clavicle *(continued)*.

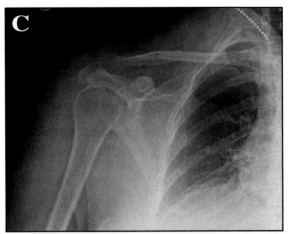

Figure 7-5 (continued). Arthroscopic distal clavicle resection. (C) Postoperative radiograph of an arthroscopic distal clavicle resection showing approximately 10 mm of resection.

Many of the postoperative rehabilitation protocols are similar, and an example of one 4-phase rehabilitation protocol starts with pain control, ROM, and isometrics exercises in phase 1, then transitions to phase 2, which focuses on isotonic strengthening exercises. Phase 3 includes increasing strength, power, endurance, and neuromuscular control, and, finally, phase 4 is dedicated to sport-specific drills.[31]

AC separations are generally treated based on the Rockwood classification system. Type I and II separations are treated with symptomatic pain relief in a sling or figure-8 brace, and early ROM exercises are initiated as soon as pain permits. These type I and II separations typically do well with non-operative treatment, and the majority of patients return to preinjury activities.[32] Patients who sustain Type I or II separations are at risk to develop post-traumatic arthritis and are treated with the same algorithm as idiopathic osteoarthritis, which includes 3 to 6 months of nonoperative treatment and distal clavicle excision for patients who proceed to operative intervention. The exceptions to this treatment are type II separations with hypermobility and type III separations, in which treatment with isolated distal clavicle resection is contraindicated. The treatment of patients with type III separations is controversial but historically has been conducted with up to a 12-week trial of nonoperative treatment. Several studies and a meta-analysis have shown no significant differences between satisfactory outcomes and strength between the operative group and the nonoperative groups.[33-36] The nonoperative groups tend to have significantly lower complication rates and earlier return to work and athletics. Most surgeons also advocate nonoperative treatment in contact athletes due to the high risk of reinjury.[19] Some authors have supported initial operative treatment in patients involved in repetitive lifting, overhead work, manual labor, and in the dominant arm in overhead athletes.[37] In one study of Type III separations in pitchers, complete pain relief and normal function were obtained in 80% of pitchers treated nonoperatively and 91% of pitchers treated operatively.[37] Operative treatment for reconstructing AC separations typically involves recreation of the CC ligaments, and procedures frequently used include the Weaver-Dunn, the modified Weaver-Dunn, anatomic CC ligament reconstruction, and other newer arthroscopic techniques.[19,38] Type IV, V, and VI separations due to significant morbidity are treated with early surgical treatment to reduce the AC joint and repair or reconstruct the disrupted soft tissues. The Weaver-Dunn procedure uses the coracoacromial ligament and transfers it to the distal clavicle to attempt to recreate the CC ligaments. However, the coracoacromial ligament has only 20% of the ultimate load of the CC ligaments, and several modifications of the Weaver-Dunn procedure have been developed to augment the coracoacromial transfer and improve the stiffness and load to failure of the reconstruction.[39] The most common modifications include the addition of a distal clavicle excision to decrease the risk of developing post-traumatic arthritis and augmentation of the transfer using CC fixation in the form of screws, sutures, and/or Mersilene tape (Ethicon, Somerville, NJ).[40-42] The anatomic reconstruction of the CC ligaments was developed in an attempt to better recreate the CC ligaments using allograft that is looped around the coracoid process and then passed through bone tunnels through the clavicle at the normal attachment sites of the conoid and trapezoid ligaments. The remaining allograft is then used to reconstruct the AC ligament and capsule (Figure 7-6).[43]

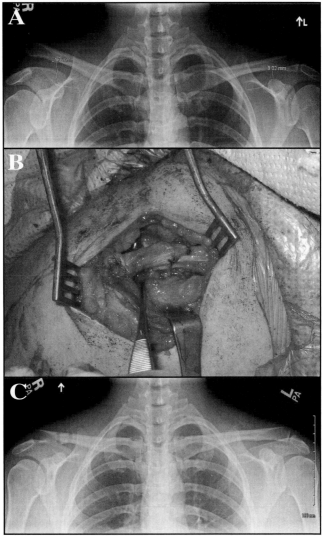

Figure 7-6. Right type V AC separation. (A) Preoperative bilateral Zanca view showing 21.3 mm of CC distance compared to 8 mm displacement on the normal side. (B) Intraoperative view of a completed anatomic reconstruction of the AC joint. (C) Postoperative bilateral Zanca view demonstrating restoration of the CC distance after anatomic CC ligament reconstruction using allograft.

Newer arthroscopic techniques have been developed to obtain and maintain reduction of the AC joint including techniques using Endobuttons (Smith & Nephew Endoscopy, Andover, MA, TightRopes (Arhtrex, Inc, Naples, FL), and a variety of suture anchor devices; however, these newer techniques, though promising, have yet to be proven with longer-term follow-up studies.[44-48]

CONCLUSION

The etiology of shoulder pain is multifactorial, and AC joint pathology is just one of many shoulder pathologies that can occur in isolation or in combination. A thorough history and physical that includes careful examination of the AC joint, supplemented by appropriate imaging, will facilitate proper diagnosis and treatment of all sources of shoulder pain and dysfunction. When examining the AC joint and attributing symptoms to it, it is important to remember that degeneration at the AC joint is exceedingly common and frequently asymptomatic. Conditions such as AC joint osteoarthritis, post-traumatic arthritis, AC separations, and distal clavicular osteolysis are the most common disorders of the AC joint and should be included in the differential diagnosis of a patient complaining of shoulder discomfort.

As with most shoulder conditions, nonoperative treatment is the first line of treatment for most AC joint conditions, and only after failure of this should open and arthroscopic treatments be used. Surgical treatment of osteoarthritis, post-traumatic arthritis, and distal clavicular osteolysis has very good and consistent results when used for the proper indications. More severe injuries to the AC joint resulting in Type IV, V, and VI separations are treated with early surgery. The treatment of Type III separations remains controversial but is typically nonoperative because surgical management has demonstrated similar results to nonoperative treatment only with higher complication rates.

REFERENCES

1. Rockwood CJ, Williams G, Young D. Disorders of the acromioclavicular joint. In: Rockwood CJ, Matsen FA III, eds. *The Shoulder.* 2nd ed. Philadelphia, PA: WB Saunders; 1998:483-553.
2. Filk K, Lyman S, Marx RG. American collegiate men's ice hockey: an analysis of injuries. *Am J Sports Med.* 2005;33:183-187.
3. Kaplan LD, Flanigan DC, Norwig J, et al. Prevalence and variances of shoulder injuries in elite collegiate football players. *Am J Sports Med.* 2005;33:1142-1146.
4. Kelly BT, Barnes RP, Powell JW, et al. Shoulder injuries to quarterbacks in the National Football League. *Am J Sports Med.* 2004;32:328-331.
5. Kocher MS, Feagin JA Jr. Shoulder injuries during alpine skiing. *Am J Sports Med.* 1996;24:665-669.
6. Shaffer BS. Painful conditions of the acromioclavicular joint. *J Am Acad Orthop Surg.* 1999;7:176-188.
7. Clark HD, McCann PD. Acromioclavicular joint injuries. *Orthop Clin North Am.* 2000;31:177-186.
8. Bergfeld JA, Andrish JT, Clancy WG. Evaluation of the acromioclavicular joint following first- and second-degree sprains. *Am J Sports Med.* 1978;6:153-159.
9. Taft TN, Wilson FC, Oglesby JW. Dislocation of the acromioclavicular joint: an end result study. *J Bone Joint Surg Am.* 1987;69:1045-1051.
10. Nuber GW, Bowen MK. Acromioclavicular joint injuries and distal clavicle fractures. *J Am Acad Orthop Surg.* 1997;5:11-18.
11. Horvath F, Kery L. Degenerative deformations of the acromioclavicular joint in the elderly. *Arch Gerontol Geriatr.* 1984;3:259-265.
12. Scavenius M, Iversen BF. Nontraumatic clavicular osteolysis in weight lifters. *Am J Sports Med.* 1992;20:463-467.
13. Cahill BR. Osteolysis of the distal part of the clavicle in male athletes. *J Bone Joint Surg Am.* 1982;64:1053-1058.
14. Chronopoulos E, Kim TK, Park HB, et al. Diagnostic value of physical tests for isolated chronic acromioclavicular lesions. *Am J Sports Med.* 2004;32(3):655-661.
15. Gerber C, Galantay RV, Hersche O. The pattern of pain produced by irritation of the acromioclavicular joint and the subacromial space. *J Shoulder Elbow Surg.* 1998;7:352-355.
16. Walton J, Mahajan S, Paxinos A, et al. Diagnostic values of tests for acromioclavicular joint pain. *J Bone Joint Surg.* 2004;86-A(4):807-812.
17. O'Brien SJ, Pagnani MJ, Fealy S, et al. The active compression test: a new and effective test for diagnosing labral tears and acromioclavicular joint abnormality. *Am J Sports Med.* 1998;26:610-613.
18. Hossain S, Jacobs LG, Hashmi R. The long-term effectiveness of steroid injections in primary acromioclavicular joint arthritis: a five-year prospective study. *J Shoulder Elbow Surg.* 2008;17:535-538.

19. Rios CG, Mazzocca AD. Acromioclavicular joint problems in athletes and new methods of management. *Clin Sports Med.* 2008;27:763-788.
20. Fukuda K, Craig EV, An KN, et al. Biomechanical study of the ligamentous system of the acromioclavicular joint. *J Bone Joint Surg Am.* 1986;68:434-440.
21. DePalma AF. The role of the disks of the sternoclavicular and acromioclavicular joints. *Clin Orthop.* 1959;13:222-233.
22. Rios CG, Arciero RA, Mazzocca AD. Anatomy of the clavicle and coracoids process for reconstruction of the coracoclavicular ligaments. *Am J Sports Med.* 2007;35(5):811-817.
23. Edelson JG. Patterns of degenerative change in the acromioclavicular joint. *J Bone Joint Surg Br.* 1996;78:242-243.
24. Zanca P. Shoulder pain: involvement of the acromioclavicular joint (analysis of 1000 cases). *AJR Am J Roentgenol.* 1971;112:493-506.
25. Bossart PJ, Joyce SM, Manaster BJ, et al. Lack of efficacy of "weighted" radiographs in diagnosing acute acromioclavicular separations. *Ann Emerg Med.* 1988;17:20-24.
26. Needell SD, Zlatkin MB, Sher JS, et al. MR imaging of the rotator cuff: peritendinous and bone abnormalities in an asymptomatic population. *AJR Am J Roentgenol.* 1996;166(4):863-867.
27. Strobel K, Pfirrmann CW, Zanetti M, et al. MRI features of the acromioclavicular joint that predict pain relief from intra-articular injection. *AJR Am J Roentgenol.* 2003;181(3):755-760.
28. Mumford EB. Acromioclavicular dislocation: a new operative treatment. *J Bone Joint Surg Am.* 1941;23:799-802.
29. Gurd FB. The treatment of complete dislocation of the outer end of the clavicle: an hitherto undescribed operation. *Ann Surg.* 1941;113:1094-1098.
30. Henry MH, Liu SH, Loffredo AJ. Arthroscopic management of the acromioclavicular joint disorder: a review. *Clin Orthop.* 1995;316:276-283.
31. Gladstone J, Wilk K, Andrews J. Non-operative treatment of acromioclavicular joint injuries. *Oper Tech Sports Med.* 1997;5:78-87.
32. Dias JJ. The conservative treatment of acromioclavicular dislocation: review after five years. *J Bone Joint Surg Br.* 1987;69(5):719-722.
33. Phillips AM, Smart C, Groom AF. Acromioclavicular dislocation: conservative or surgical therapy. *Clin Orthop.* 1998;353:10-17.
34. Galpin RD, Hawkins RJ, Grainger RW. A comparative analysis of operative versus non-operative treatment of grade III acromioclavicular separations. *Clin Orthop.* 1985;193:150-155.
35. Press J, Zuckerman JD, Gallagher M, et al. Treatment of grade III acromioclavicular separations: operative versus non-operative management. *Bull Hosp Joint Dis.* 1997;56:77-83.
36. Tibone J, Sellers R, Tonino P. Strength testing after third degree acromioclavicular dislocations. *Am J Sports Med.* 1992;20:328-331.
37. McFarland EG, Blivin SJ, Doehring CB, et al. Treatment of grade III acromioclavicular separations in professional throwing athletes: results of a survey. *Am J Orthop.* 1997;26:771-774.

38. Weaver JK, Dunn HK. Treatment of acromioclavicular injuries, especially complete acromioclavicular separation. *J Bone Joint Surg Am*. 1972;54:1187-1194.

39. Motamedi AR, Blevins FT, Willis MC, et al. Biomechanics of the coracoclavicular ligament complex and augmentations used in its repair and reconstruction. *Am J Sports Med*. 2000;28:380-384.

40. Tienen TG, Oyen JF, Eggen P. A modified technique for complete acromioclavicular dislocation: a prospective study. *Clin Orthop*. 2003;31(5):655-659.

41. Rokito AS, Oh YH, Zuckerman JD. Modified Weaver-Dunn procedure for acromioclavicular joint dislocations. *Orthopedics*. 2004;27:21-28.

42. Rolla PR, Surace MF, Murena L. Arthroscopic treatment of acute acromioclavicular joint dislocation. *Arthroscopy*. 2004;20:662-668.

43. Mazzocca AD, Conway JE, Johnson SJ, et al. The anatomic coracoclavicular reconstruction. *Oper Tech Sports Med*. 2004;12:56-61.

44. Clavert P, Moulinoux P, Kempf J. Technique of stabilization in acromioclavicular joint dislocation. *Tech Shoulder Elbow Surg*. 2005;6:1-7.

45. Chernchujit B, Tischer T, Imhoff AB. Arthroscopic reconstruction of the acromioclavicular joint disruption: surgical technique and preliminary results. *Arch Orthop Trauma Surg*. 2006;126:575-581.

46. Struhl S. Double Endobutton technique for repair of complete acromioclavicular joint dislocations. *Tech Shoulder Elbow Surg*. 2007;8:175-179.

47. Wellmann M, Zantop T, Petersen W. Minimally invasive coracoclavicular ligament augmentation with a flip button/polydioxanone repair for treatment of total acromioclavicular joint dislocation. *Arthroscopy*. 2007;23:1132.

48. Lim YW, Sood A, van Riet RP, et al. Acromioclavicular joint reduction, repair, and reconstruction using metallic buttons—early results and complications. *Tech Shoulder Elbow Surg*. 2007;8:213-221.

8

GLENOHUMERAL ARTHRITIS

Gerald R. Williams Jr, MD

INTRODUCTION

Glenohumeral (GH) arthritis may be defined as loss of articular cartilage on the humerus and/or glenoid associated with varying degrees of bone loss, capsular contracture, and rotator cuff insufficiency, depending on the underlying cause of the arthritis. Common types of GH arthritis include osteoarthritis, avascular necrosis, inflammatory arthritis, crystalline arthritis, post-traumatic arthritis, arthritis of instability, rotator cuff tear arthropathy, neuropathic arthropathy, and septic arthritis (Table 8-1).

The history, physical examination, and radiographic findings exhibited by patients with GH arthritis generally include

Cohen SB. *Musculoskeletal Examination of the Shoulder:*
Making the Complex Simple (pp. 161-195).
© 2011 SLACK Incorporated

Table 8-1

TYPES OF GLENOHUMERAL ARTHRITIS

Primary Osteoarthritis
  Primary
  Secondary
    Post-traumatic
    Instability

Inflammatory Arthritis
  Rheumatoid arthritis
  Ankylosing spondylitis
  Psoriatic arthritis
  Inflammatory bowel disease

Cuff Tear Arthropathy

Crystalline Arthropathy
  Gout
  Pseudo-gout

Osteonecrosis
  Idiopathic
  Corticosteroid-induced
  Alcoholism
  Post-traumatic
  Gaucher's disease
  Post-irradiation necrosis
  Hemoglobinopathies
    Sickle cell disease
    Hemophilia
    Hemachromatosis

Neuropathic Arthropathy
  Syringomyelia
  Peripheral neuropathy

Septic Arthritis

Lyme Arthritis

Arthritis Associated with Acromegaly

Congenital
  GH dysplasia

some degree of insidious pain and stiffness associated with progressive loss of radiographic joint space. However, findings vary considerably depending on the relative involvement of the underlying bone, capsule, and rotator cuff. Rotator cuff integrity is one key factor in determining physical findings and prognosis following surgical treatment. The various types of arthritis comprise a spectrum of rotator cuff involvement with avascular necrosis and primary osteoarthritis on 1 end (0% to 10% rotator cuff tears), cuff tear arthropathy (100% rotator cuff tears) on the other, and all others somewhere in between.[1-3] Rheumatoid arthritis is the prototypical form of inflammatory arthritis and has been reported to have 20% to 30% full-thickness rotator cuff tears.[4] For the purposes of this chapter, discussion will be focused on findings associated with primary osteoarthritis, rheumatoid arthritis, and cuff tear arthropathy.

HISTORY

In general, patients with any type of GH arthritis will report intermittent pain for a long time, often for many years. It is usually worse with activities and may be associated with mechanical popping and clicking. Often, the symptoms will have become worse over recent months. Depending on the degree of involvement, the pain will be associated with a perceptible loss of motion. Historical features may vary slightly according to the underlying type of arthritis and are summarized in Table 8-2.

Osteoarthritis

The hallmarks of primary osteoarthritis of the GH joint include a dull, achy type of pain that typically has been present for many years. It is often worse with activity and weather changes. However, paradoxically, patients will often note that some level of activity is beneficial. Night pain is ubiquitous and can interfere substantially with sleep. Sleep interference is often the prime factor that prompts medical attention. Symptoms frequently will be bilateral but asymmetric, with one side (often the dominant one) being more severe. There may be a remote history of trauma. Involvement of other joints, such as the hip or knee, is quite common.

Table 8-2

HISTORY AND PHYSICAL FINDINGS—OSTEOARTHRITIS, RHEUMATOID ARTHRITIS, CUFF TEAR ARTHROPATHY

	History	Inspection	Palpation	Range of Motion	Strength	Neurovascular
Osteoarthritis	Dull, achy pain; Worse with activity; Night pain; Mechanical popping; Other joint involvement (Heberden's and Bouchard's nodes); Male preponderance	Generalized atrophy (disuse); Occasional AC joint prominence; Posterior prominence of the humeral head (posterior subluxation); Anterior prominence of the acromion (posterior subluxation)	Posterior GH joint line tenderness; AC joint tenderness (when AC joint symptomatic)	May be difficult to test because of pain; Generally, active and passive range of motion decreased equally; Motion loss global but worse in external rotation	May be unreliable because of pain; Generally good	Usually intact

(continued)

Table 8-2 (continued)

HISTORY AND PHYSICAL FINDINGS—OSTEOARTHRITIS, RHEUMATOID ARTHRITIS, CUFF TEAR ARTHROPATHY

	History	Inspection	Palpation	Range of Motion	Strength	Neurovascular
Rheumatoid arthritis	Early morning stiffness and pain Multiple joint involvement, symmetrical Relatively young (3rd and 4th decade) Female preponderance	Often thin, friable skin Generalized atrophy (disuse and systemic involvement) Spinatus atrophy (cases of cuff tear) AC joint prominence	AC joint tenderness when symptomatic Dorsal acromion (stress fracture)	Severe stiffness can mask cuff insufficiency Active motion often less than passive ER lag sign (posteriosuperior cuff tear) IR lag sign (subscapularis tear)	May be unreliable because of pain ER weakness at 0 degrees (supra- and infraspinatus rupture) ER weakness at 90 degrees (teres minor rupture) IR weakness (subscapularis rupture)	Usually normal C-spine instability screening Preoperative baseline, especially if on methotrexate

(continued)

Table 8-2 (continued)

History and Physical Findings—Osteoarthritis, Rheumatoid Arthritis, Cuff Tear Arthropathy

	History	Inspection	Palpation	Range of Motion	Strength	Neurovascular
Cuff tear arthropathy	Elderly presentation (7th or 8th decade)	Supra- and infraspinatus atrophy	AC joint tenderness when symptomatic	Active motion usually less than passive motion	Marked variability	Usually intact
	Female preponderance	AC joint cyst	Dorsal acromial tenderness (os acromiale, stress fracture/nonunion)	Lag signs frequent Active	Usually some degree of ER weakness	Check deltoid (prior surgery)
	Bilateral	Large subdeltoid or subcutaneous fluid collection (often hemorrhagic)		Elevation can be remarkably well preserved, especially if no surgery or trauma	ER weakness at 0 degrees poor prognostic factor (teres minor involvement)	
	History of "bursitis" or "tendonitis"	Surgical scars (deltoid detachment)				
	Multiple effusions, often blood streaked					

The fourth or fifth decade of life is the most common presenting age, and there is a male preponderance of approximately
1.5 to 2:1.[5,6]

Rheumatoid Arthritis

New onset of rheumatoid arthritis in any joint may be
acutely more painful than primary osteoarthritis. Rheumatoid
arthritis represents a marked inflammatory response with
acute synovitis. Rarely is the shoulder the primary presenting
joint. By the time patients present with symptomatic rheumatoid arthritis of the GH joint, the diagnosis has usually already
been made, and treatment has often already been instituted.
The common complaints are pain and stiffness that is often
worse in the morning and improves throughout the day as the
joints become more active. Bilateral, nearly symmetric involvement is common, and other joints are usually affected, again
usually in a symmetric fashion. Ipsilateral upper extremity
involvement is important to elicit as it may have a substantial
impact on overall function. Rheumatoid patients are typically
young (3rd and 4th decade of life) at the time of presentation
and have a 2 to 3:1 female preponderance.[7]

Special consideration should be given to the presence of
neck pain and associated numbness and paresthesias in the
rheumatoid patient with shoulder pain. Segmental instability
and other problems of the cervical spine are common in rheumatoid arthritis and can be lifethreatening.

Cuff Tear Arthropathy

Cuff tear arthropathy, as described by Neer et al, typically presents in an elderly (7th or 8th decade) female (3 to
4:1 female:male preponderance) with repeated episodes of
substantial shoulder swelling or fluid accumulation, pain,
weakness, discoloration or ecchymosis, and difficulty rais-
ing the arm.[3] Symptoms have usually been present for many
years, often with well-defined periods of exacerbation. It is
often bilateral, and patients have frequently been treated for
"bursitis" or "tendonitis" or have been told in the past that
they had a "tear." A history of drainage, fever, or chills should
be elicited, as occasionally infection from a corticosteroid
injection may be encountered.

GH arthritis in the presence of irreparable rotator cuff insufficiency may also occur without the classic history described by Neer et al.[3] These patients are more often male and younger and have frequently had prior surgery for rotator cuff repair. Consequently, symptoms of coracoacromial arch insufficiency, such as anterosuperior escape and pseudoparalysis, may be more common. The possibility of infection should also be excluded as much as possible.

Dividing patients with irreparable rotator cuff insufficiency and GH arthritis into the above 2 categories is obviously an oversimplification. The key point here is that the presentation of these patients is extremely heterogeneous, with some patients reporting minimal pain, good overhead function, and minimal weakness, and others complaining of severe pain, recurrent swelling, and pseudoparalysis and everything in between.

EXAMINATION

Physical examination techniques used in the evaluation of patients suspected of having GH arthritis include inspection, palpation, range of motion testing, strength testing, and neurovascular testing. Although some findings will be consistent across the entire spectrum of arthritis, certain findings are more prominent in specific types of arthritis. It is important to inspect the entire shoulder girdle unclothed and to inspect the joints above (ie, cervical spine) and below the shoulder. Physical findings are summarized in Table 8-3.

Osteoarthritis

Inspection of the shoulder girdle in patients with long-standing osteoarthritis will reveal generalized atrophy of the entire shoulder secondary to disuse. Flattening of the anterior deltoid and generalized loss of muscle mass of a mild to moderate degree are often present. However, it is uncommon to see specific atrophy of the supra- and infraspinatus because full-thickness rotator cuff tears are unusual and generally small. The acromioclavicular (AC) joint may appear prominent. In addition, prominence of the medial border of the

Table 8-3

HELPFUL HINTS

Type of Arthritis	History	Examination Findings	Radiographic Findings
Osteo-arthritis	Dull achy pain, changes in weather, night pain	Deltoid atrophy, AC joint hypertrophy, symmetric passive and active ROM loss (particularly external rotation), strength usually normal	Asymmetric joint space narrowing, subchondral sclerosis, osteophyte formation, and subchondral cyst formation. Osteophyte formation is most obvious along the inferior humeral neck. The glenoid is worn posteriorly and may be associated with posterior subluxation
Inflammatory arthritis (eg, rheumatoid)	Pain and morning stiffness, other joint involvement (rare for shoulder to present first), bilateral and symmetric involvement	Upper extremity deformity, general shoulder girdle atrophy, active ROM < passive ROM, weakness with associated cuff tears	Regional osteopenia, symmetrical joint space narrowing, and juxta-articular erosions. Erosions are best seen at the synovial reflection on the superior aspect of the humeral head. Central glenoid erosions. MRI scans are helpful for glenoid and humeral erosions and also rotator cuff tears
Rotator cuff arthropathy	Elderly patients, shoulder swelling, chronic symptoms with exacerbations, usually a prior history of a rotator cuff tear, can be younger male with prior rotator cuff repair	Supraspinatus and/or infraspinatus atrophy, effusion, active ROM < passive ROM, possible anterosuperior escape, rotator cuff weakness	Acromiohumeral space narrowing with superior migration of humeral head. Humeral head collapse. MRI scans will show a chronic rotator cuff tear with muscle atrophy

scapula may be present, but rarely is it on a neurological basis. Prominence or fullness in the posterior aspect of the shoulder may represent posterior humeral subluxation and can often be appreciated in thin individuals. Although involvement of the remaining upper extremity joints is uncommon, painful swelling of the distal interphalangeal joints (Heberden's nodes) and proximal interphalangeal joints (Bouchard's nodes) may be seen, especially in patients with generalized osteoarthritis involving multiple joints.[8]

Many areas of the shoulder girdle are subcutaneous and amenable to palpation. In patients with primary osteoarthritis, the 2 most important areas for palpation are the AC joint and the posterior GH joint line. AC joint arthritis may accompany GH arthritis and can be identified initially by pain with palpation. The most common area of tenderness to palpation in GH osteoarthritis is the posterior joint line. This is particularly true in patients with posterior subluxation.[5]

When testing range of motion, both the active and passive range of motion should be noted. Primary GH osteoarthritis is characterized by symmetric loss of both active and passive range of motion. Although pain may interfere with motion measurement, with persistence, the examiner can most often demonstrate equal active and passive range of motion in most patients with primary osteoarthritis.

The motion loss in primary osteoarthritis occurs in all planes. However, it is most pronounced in external rotation with the arm at the side and with the arm at 90 degrees of elevation in the scapular plane—assuming that 90 degrees of elevation can be achieved (Figure 8-1).

Strength testing is primarily performed to verify suspected rotator cuff tears. It can be unreliable, especially if the shoulder is very stiff and painful. In patients with osteoarthritis, external and internal rotation strength with the arm at the side is generally preserved.

Neurovascular examination in patients with primary osteoarthritis is usually normal. The one potential exception is female patients who have undergone mastectomy. Although arterial, venous, and neurological function are usually normal, abnormal lymph flow can be an issue after surgery. Any sign of edema or swelling in the hand may represent subclinical lymphedema and should be noted.

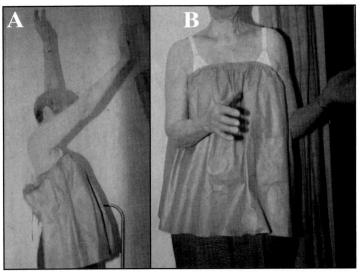

Figure 8-1. Osteoarthritis is characterized by similar loss of active and passive motion. (A) Loss of motion includes all planes and (B) is more severe in external rotation.

Rheumatoid Arthritis

Although the degree of involvement of upper extremity joints is variable among patients with rheumatoid arthritis, most patients who present with symptomatic GH rheumatoid arthritis will have obvious deformity of the hands (especially the metacarpophalangeal joints), wrists, and elbows. The skin may be thin and friable. Muscles of the entire shoulder girdle are atrophic, not only because of disuse but also because of the systemic effects of the disease. Moreover, many patients will demonstrate specific atrophy of the supra- and/or infraspinatus muscles because of rotator cuff tears (Figure 8-2). The AC joint is often involved and may be prominent to visual inspection.

Palpation should be carried out at the level of the AC joint as well as the dorsal aspect of the acromion. Because the AC joint is a synovial joint, it too can be involved with rheumatoid arthritis. This can be suspected initially on the basis of tenderness. In addition, because of the osteoporosis associated with rheumatoid arthritis, combined with proximal humeral

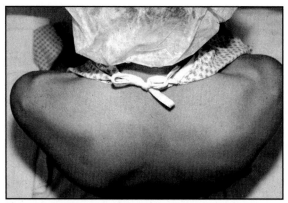

Figure 8-2. Note atrophy in the supra- and infraspinatus seen in long-standing posterosuperior rotator cuff tears.

migration because of chronic rotator cuff insufficiency, stress fractures of the acromion can occur. Again, this can be suspected on the basis of point tenderness over the dorsal aspect of the acromion, usually in its mid-portion directly lateral to the posterior aspect of the AC joint.

Unless the shoulder is severely stiff (ie, passive elevation less than 90 degrees), active range of motion is often less than passive range of motion in patients with rheumatoid arthritis (Figure 8-3). This is usually because of full-thickness rotator cuff tears or partial tearing and associated rotator cuff dysfunction. When maximum passive external rotation is greater than active external rotation, the difference is termed an *external rotation lag sign*.[9] When performed with the arm at the side, the presence of an external rotation lag sign is indicative of insufficiency of the supraspinatus and upper infraspinatus (Figure 8-4). When present with the arm at 90 degrees of elevation in the scapular plane, the external rotation lag sign is indicative of teres minor and lower infraspinatus tearing or dysfunction (Figure 8-5).[10]

Lag signs for internal rotation range of motion can also be observed and suggest subscapularis insufficiency.[9] One version of the internal rotation lag sign involves placing the back of the patient's hand against his or her sacrum and then passively internally rotating the humerus so that the hand moves away from the sacrum. A positive internal rotation lag sign

Figure 8-3. Rheumatoid arthritis is character-ized by symmetrical loss of motion, often with more passive than active range of motion. Note involvement of all upper extremity joints.

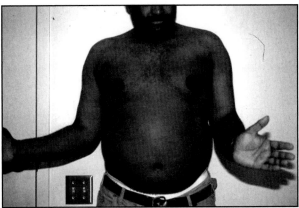

Figure 8-4. This patient has equal passive external rotation in both shoulders but cannot actively maintain his passive external rotation because of a rotator cuff tear involving the supraspinatus and upper infraspinatus. This is termed an *external rotation lag sign*.

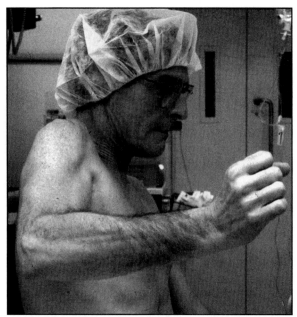

Figure 8-5. This patient with a rotator cuff tear involving the lower infraspinatus and teres minor is unable to actively rotate his elevated arm to maximum passive external rotation. This is also termed an *external rotation lag sign*.

occurs when the patient cannot hold the hand in this maximally internally rotated position. Many patients with rheumatoid arthritis cannot place their hand in this position because of pain or lack of passive motion. Under these circumstances, the palm is placed on the abdomen and held there while the ipsilateral elbow is pulled anteriorly to internally rotate the humerus. An internal lag sign is present when the patient cannot hold the elbow forward without allowing the palm to come away from the abdomen (Figure 8-6).

External and internal rotation strength can be combined with information concerning lag signs to get an impression of rotator cuff strength. For example, if the external rotation lag sign is greater and the strength is less with the arm at the side than with the arm at 90 degrees of elevation, the rotator cuff is likely deficient in the region of the supraspinatus and upper portion of infraspinatus, with an intact inferior infraspinatus

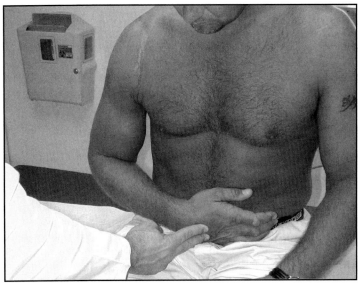

Figure 8-6. This patient is unable to actively maintain maximal passive humeral internal rotation (elbow forward) because of a subscapularis tear. This is termed an *internal rotation lag sign.*

and teres minor. Pain and stiffness may prevent accurate assessment of strength as with osteoarthritis.

Neurovascular examination in patients with rheumatoid arthritis is usually normal but should be documented. Rheumatoid patients who take methotrexate may be at increased risk of brachial plexus injury following arthroplasty, and a baseline exam is useful.[11]

Cuff Tear Arthropathy

Without exception, inspection of the shoulder in patients with cuff tear arthropathy or GH arthritis associated with a large rotator cuff tear reveals atrophy of the supraspinatus and, usually, the infraspinatus muscles. The presence of fluid under the deltoid and, occasionally, subcutaneously over the AC joint can be visualized. These fluid collections can sometimes be very large and contain blood that can be seen subcutaneously as ecchymotic areas. Surgical scars can be observed in patients who have undergone prior rotator cuff

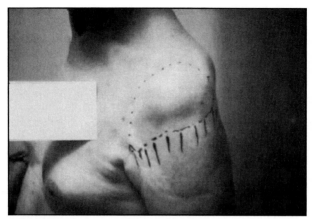

Figure 8-7. Cuff tear arthropathy can be associated with deltoid detachment in patients who have had prior surgery. Surgical incisions (dotted line) and deltoid detachment with retraction (dashed line) should be noted.

repair. Visual evidence of dehiscence of the anterior or middle deltoid attachments should be noted (Figure 8-7). Although deltoid detachment is more commonly seen postsurgically, it can also occur spontaneously in patients with massive, long-standing rotator cuff tears.[12] The remaining joints of the upper extremity are usually normal.

The acromion and AC joint should be palpated for tenderness. The AC joint may be tender in patients with concomitant AC arthritis. In addition, tender os acromiales or acromial stress fracture nonunions may be identified by palpation. Finally, postsurgical or spontaneous deltoid detachment may be associated with tenderness and prominence of the "naked" anterior acromion.

Range of motion testing can be very revealing in patients with cuff tear arthropathy or GH arthritis with irreparable cuff insufficiency. Just like in rheumatoid arthritis, in the absence of severe stiffness, active range of motion will be less than passive range of motion. This difference may not be as noticeable in some planes compared to others. Some of these patients will have remarkably well-preserved elevation despite having large or massive rotator cuff tears. This is particularly true in the absence of prior rotator cuff surgery or trauma. However, external and/or internal rotation lag signs of varying degrees are almost always present. Conversely, patients with relatively minor rotational lag signs may have pseudoparalysis and anterosuperior escape in the presence of postsurgical deltoid detachment or coracoacromial arch deficiency. Strength testing in patients with cuff tear arthropathy can provide similar information regarding the rotator cuff as can be garnered from strength testing in osteoarthritis and rheumatoid arthritis. One additional piece of information can be helpful when deciding upon treatment options. Improvement in function following shoulder arthroplasty, and likely also following nonoperative management, is related to whether the remaining rotator cuff can sufficiently prevent superior migration of the humerus during elevation so that GH rotation can occur and the arm can be elevated. Therefore, the ability of the patient to raise his or her arm overhead and the strength with which he or she can do it is predictive of whether hemiarthroplasty or nonoperative treatment alone can improve function. Therefore, in addition to quantifying rotational strength, the examiner should also estimate elevation strength.

Neurovascular examination in patients with cuff tear arthropathy is usually normal. However, in patients with GH arthritis and irreparable rotator cuff tears who have had prior surgery, firing of all 3 portions of the deltoid should be verified.

PATHOANATOMY

The pathoanatomy of all forms of GH arthritis can be elucidated on the basis of variable involvement of the supporting bone and soft tissues—the capsule and rotator cuff. As mentioned previously, this spectrum of disease entities can be explained using primary osteoarthritis, rheumatoid arthritis, and cuff tear arthropathy as examples.

Osteoarthritis

The pathoanatomic findings in any joint with osteoarthritis are asymmetric joint space narrowing, osteophyte formation, asymmetric soft tissue contracture, and, in severe cases, fixed joint subluxation. The shoulder is no different. The asymmetric wear occurs posteriorly, especially posteroinferiorly.[2,5] The joint subluxes posteriorly, often independent of glenoid wear. Posterior subluxation is often associated with contracture of the anterior joint capsule and, in severe cases, the subscapularis. Osteophytes form on both sides of the joint but most severely on the humerus. This causes enlargement and flattening of the humeral head that, in turn, progressively restricts motion. The rotator cuff is intact (ie, no full-thickness tears) in 90% to 95% of cases.[2,6] The long head of the biceps is pathologic in a high percentage of cases.

One last factor worth mentioning is congenital hypoplasia of the glenoid. The influence of congenital glenoid hypoplasia on the development of osteoarthritis is not known. However, there is some suggestion that patients with hypoplastic glenoids have a higher prevalence of GH arthritis than patients without glenoid hypoplasia.[13]

Rheumatoid Arthritis

Early articular changes associated with rheumatoid arthritis include damage to the microvasculature, synovial edema, and proliferation of synovial lining cells. The proliferating synovium invades the joint space in the form of villous projections. This hypertrophied synovium undergoes neovascularization and, with inflammatory cells and granulation tissue, forms pannus. This pannus destroys the articular cartilage and bone beginning at the synovial reflection at the anatomical neck of the humerus, extending onto the articular surfaces, and through to the subchondral bone. The 2 major pathoanatomical characteristics associated with rheumatoid arthritis in the shoulder are bone destruction (ie, erosion) and rotator cuff insufficiency. The bone erosion on the glenoid side is relatively concentric early on in the process and moves to the superior glenoid as rotator cuff insufficiency progresses. On the humeral side, erosion occurs around the periphery of the articular surface near the rotator cuff insertion. This juxta-articular erosion undoubtedly contributes to rotator cuff insufficiency.

Cuff Tear Arthropathy

The cause of cuff tear arthropathy is unknown, and the 2 most common theories are based on biologic and mechanical factors, respectively.[3,14] Despite the controversy surrounding the etiology of cuff tear arthropathy, the pathoanatomic features are well-accepted. They include massive, irreparable rotator cuff tears, proximal humeral head migration, recurrent blood-streaked effusions, biceps tendon dislocation and rupture, acromial nonunions, GH instability, humeral head collapse, and variable glenoid erosion—usually anterosuperior.

As mentioned previously, there is great heterogeneity in the pathoanatomy of this group of patients. Neer et al stated that humeral head collapse was a requirement for cuff tear arthropathy.[3] However, evidence supports the concept that chronic rotator cuff insufficiency leads to progressive GH arthritis, even if humeral head collapse does not occur.[15]

IMAGING

Imaging is critical in the evaluation of GH arthritis (Table 8-4). In reality, GH arthritis is primarily a radiographic diagnosis. Imaging modalities that have a role in the diagnosis of GH arthritis include plain radiography, computed tomographic (CT) scanning, and magnetic resonance image (MRI) scanning. Plain radiography is helpful in all types of arthritis and should be obtained in all patients. Standard projections include anteroposterior views in internal and external rotation, an axillary view, and a trans-scapular Y view. CT and MRI scanning have specific indications in certain types of arthritis. Radiographic findings are specific to individual diagnoses.

Osteoarthritis

Cardinal radiographic features of primary osteoarthritis include asymmetric joint space narrowing, subchondral sclerosis, osteophyte formation, and subchondral cyst formation. Osteophyte formation is most obvious along the inferior humeral neck but is present circumferentially. The glenoid is worn posteriorly and may be associated with posterior subluxation (Figure 8-8). Walch and colleagues have classified

Table 8-4

IMAGING FINDINGS—OSTEOARTHRITIS, RHEUMATOID ARTHRITIS, CUFF TEAR ARTHROPATHY

	Radiographs	CT Scan	MRI Scan
Osteo-arthritis	Subchondral sclerosis Subchondral cysts Osteophyte formation (especially inferior humerus) Asymmetric posterior wear	Routinely used to quantify posterior glenoid bone loss	Uncommonly used, rotator cuff usually intact (90% to 95%) Can be used for quantifying glenoid bone loss, but CT is better
Rheumatoid arthritis	Regional osteopenia Symmetric joint space narrowing (except with late rotator cuff insufficiency) Humeral erosion (juxta-articular at humeral neck) Glenoid erosion (central early, superior late with cuff deficiency)	Because of soft-tissue involvement, MRI is a better choice CT can show bone erosion, available glenoid bone stock	Often used to evaluate for cuff deficiency (20% to 30% full-thickness tears) Can also show bone erosion
Cuff tear arthropathy	Proximal humeral migration Humeral head collapse Usually scant osteophyte formation Superior glenoid erosion Scalloping of the coracoacromial arch ("acetabularization") Smoothing of greater tuberosity (femoralization) Acromial stress fracture/os acromiale	Can show superior glenoid erosion MRI more useful as it also quantifies rotator cuff tear	Useful for demonstrating extent of rotator cuff involvement (especially teres minor) Also demonstrates glenoid erosion

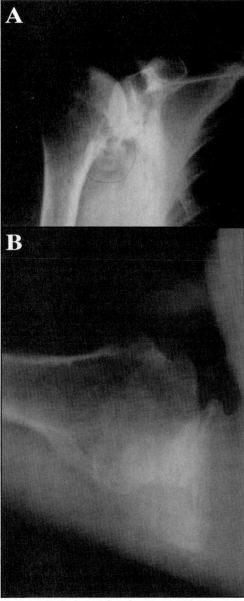

Figure 8-8. Radiographs of patients with (A) osteo-arthritis reveal subchondral sclerosis and large humeral osteophytes and (B) asymmetric joint space narrowing with posterior subluxation.

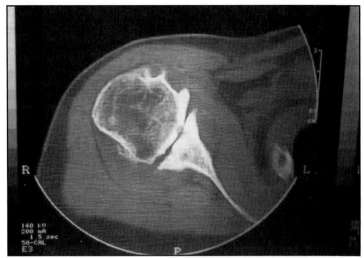

Figure 8-9. CT scanning from this patient with osteoarthritis reveals not only subchondral sclerosis and cyst formation but also asymmetrical posterior wear and mild posterior subluxation.

glenoid morphology in osteoarthritis according to the presence of posterior subluxation and wear.[16] Type A glenoids are concentric; type B glenoids are posteriorly subluxated (B1) or posterior subluxated and posteriorly worn (B2); type C glenoids are hypoplastic with increased glenoid retroversion.[16] Although posterior wear and subluxation can be identified on an axillary view, CT scanning may provide a more accurate estimate of posterior glenoid bone loss (Figure 8-9).

Rheumatoid Arthritis

The radiographic features of rheumatoid arthritis in the shoulder parallel those of any joint affected by rheumatoid arthritis. Cardinal radiographic features include regional osteopenia, symmetrical joint space narrowing, and juxta-articular erosions. These erosions are best seen at the synovial reflection on the superior aspect of the humeral head on the anteroposterior views. Osteophytes are not a prominent feature, but may be present. The anteroposterior views will also demonstrate central glenoid erosion and, in the case of rotator cuff insufficiency, superior migration of the humerus

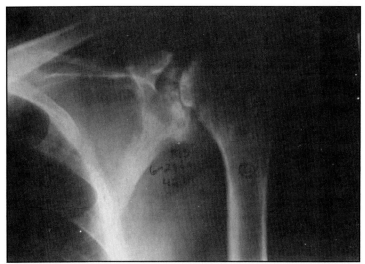

Figure 8-10. This anteroposterior radiograph of a patient with rheumatoid arthritis reveals regional osteopenia, proximal humeral migration, and superocentral erosion of the glenoid.

and possible superior glenoid erosion (Figure 8-10). The axillary view is the best way to estimate joint space narrowing, medial erosion of the glenoid surface with respect to the coracoid, and the extent of humeral head erosion. MRI scans are helpful in rheumatoid patients because they elucidate not only glenoid and humeral erosions but also rotator cuff tears (Figure 8-11).

Cuomo et al, based on observations from Neer, recognized different types of rheumatoid arthritis of the shoulder based upon radiographic appearance and grouped them into "dry," "wet," and "resorptive" patterns of involvement.[17] Each group was subdivided into low, intermediate, or severe change. In the dry type, there is loss of joint space, sclerosis, subchondral cysts, small osteophytes, and stiffness. In the wet type, there is significant synovial inflammation, marginal erosions, and medial migration of the humeral head into the glenoid. End-stage bony degenerative changes and significant bone resorption characterize the last pattern of involvement, the resorptive type.

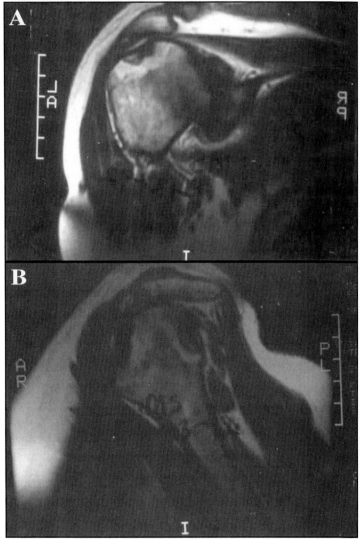

Figure 8-11. MRI scanning of the patient from Figure 8-10 reveals (A) extreme thinning of the supraspinatus tendon and proximal humeral migration but also (B) severe juxta-articular erosions of the humerus.

TREATMENT

Conservative

All patients with GH arthritis should exhaust nonoperative management before considering surgical replacement. Nonoperative treatment modalities include activity modification, nonsteroidal anti-inflammatory medications, intra-articular corticosteroid injections, and limited physiotherapy. The use of intra-articular hyaluronic acid derivatives is currently not approved for use in the shoulder by the US Food and Drug Administration (FDA) but may be a useful modality in the future. In patients with rheumatoid arthritis, consultation with a rheumatologist for the possible use of newer remitive agents is indicated. Although passive stretching exercises to maximize range of motion may decrease pain, excessive physiotherapy may increase pain. Therefore, progress should be carefully monitored, and exercises should be modified or terminated if progressive pain occurs.

Surgical

Surgical options are considered when pain and dysfunction justify surgical intervention, nonoperative management has failed, medical comorbidities do not preclude surgery, and the patient is willing to accept the risks of surgery and the responsibility of postoperative rehabilitation and activity limitations. In general, procedures can be divided into joint sparing and joint sacrificing (Table 8-5). Factors important in selecting the most appropriate procedure include age, activity level, arthritis severity, rotator cuff integrity, and bone deformity. Joint-sparing procedures include open or arthroscopic débridement; synovectomy; and/or capsular release, open or arthroscopic soft tissue interposition, and humeral or glenoid osteotomy. Joint-sacrificing procedures include arthrodesis, resection arthroplasty, and joint replacement.

A detailed discussion of surgical indications is beyond the scope of this chapter. However, as a rule, joint-sparing procedures are reserved for younger (under 40), more active patients, especially if their disease is less severe, and joint-sacrificing procedures are indicated when the joint is

Table 8-5

SURGICAL OPTIONS FOR GLENOHUMERAL ARTHRITIS

Joint Sparing	Joint Sacrificing
Soft-tissue release, débridement, synovectomy • Arthroscopic • Open	Arthrodesis
Soft-tissue interposition • Arthroscopic • Open	Resection arthroplasty

Osteotomy
• Glenoid
• Humeral
	Joint replacement
	• Humeral resurfacing (± biologic glenoid resurfacing)
	• Humeral replacement (± biologic glenoid resurfacing)
	• Total shoulder replacement
	  o Anatomic
	  o Reverse

too severely involved to be saved. Arthrodesis and resection arthroplasty are rarely performed for primary arthritis. The primary indications for arthrodesis in the management of GH arthritis are combined deltoid and rotator cuff loss or salvage of a failed prosthesis. Rarely, arthrodesis may be required in a very young manual laborer. Resection arthroplasty is usually only indicated as a salvage procedure in infectious arthritis or a failed prosthesis.

Joint replacement is the most commonly performed joint-sacrificing procedure and provides predictable pain relief and improved function in the vast majority of cases. Options include humeral head resurfacing, humeral head replacement, total shoulder replacement, and reverse total shoulder replacement. Indications for each type of joint replacement vary with rotator cuff integrity, glenoid involvement, patient age and activity level, and surgeon preference.

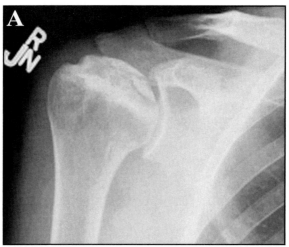

Figure 8-12. Anteroposterior radiograph of a young, active patient with (A, B) GH arthritis and a concentric glenoid who was treated with (C) humeral resurfacing arthroplasty *(continued).*

In general, humeral resurfacing is indicated in patients who are not candidates for glenoid replacement. Glenoid reaming and replacement can be done with resurfacing-type implants, but exposure is much more difficult than if the head is removed. Typical patients undergoing humeral resurfacing include those with avascular necrosis and minimal glenoid involvement and those with osteoarthritis who are young (under 50) and have minimal glenoid deformity (Figure 8-12). Humeral head replacement is performed when metaphyseal bone is inadequate for fixation of a resurfacing implant or the glenoid requires substantial reaming or biologic resurfacing (Figure 8-13). Guidelines for total shoulder arthroplasty include age older than 50, humeral and glenoid involvement, and an intact or reparable rotator cuff (Figure 8-14). Reverse total shoulder replacement is accompanied by a higher complication rate than anatomic shoulder replacement. Therefore, it is reserved for older (70 or older), more sedentary patients with GH arthritis combined with irreparable rotator cuff insufficiency (Figure 8-15).

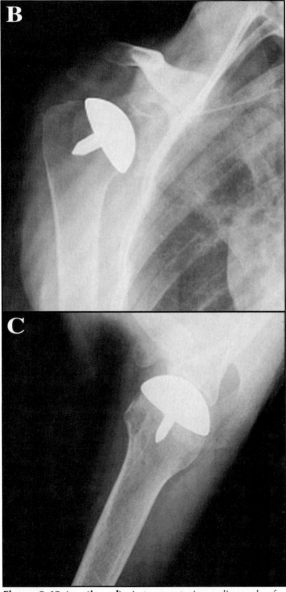

Figure 8-12 (continued). Anteroposterior radiograph of a young, active patient with (A, B) GH arthritis and a concentric glenoid who was treated with (C) humeral resurfacing arthroplasty.

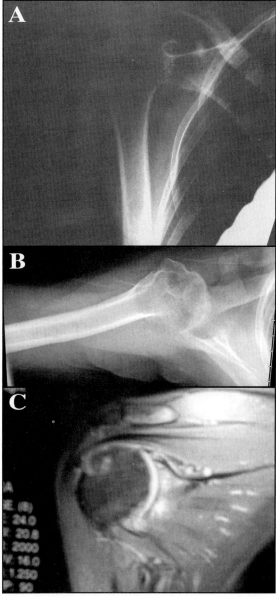

Figure 8-13. This young (27 year old) patient with (A, B) juvenile rheumatoid arthritis and an intact rotator cuff was treated with (C through F) humeral head replacement and biologic glenoid resurfacing with a meniscal allograft *(continued).*

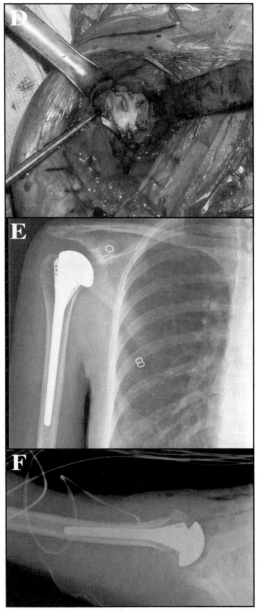

Figure 8-13 (continued). This young (27 year old) patient with (A, B) juvenile rheumatoid arthritis and an intact rotator cuff was treated with (C through F) humeral head replacement and biologic glenoid resurfacing with a meniscal allograft.

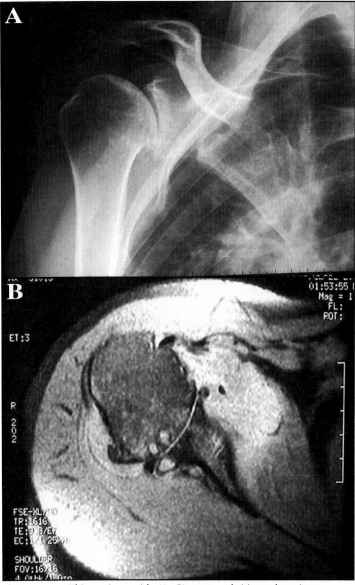

Figure 8-14. This patient with (A, B) osteoarthritis and an intact rotator cuff was treated with (C, D) anatomic total shoulder arthroplasty *(continued).*

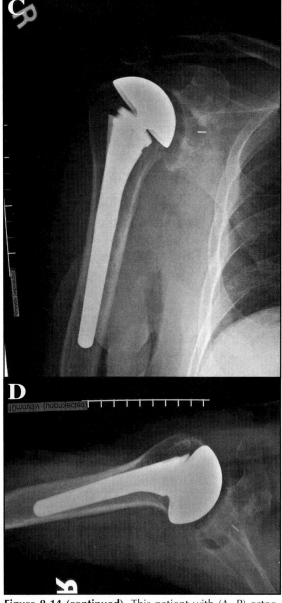

Figure 8-14 (continued). This patient with (A, B) osteoarthritis and an intact rotator cuff was treated with (C, D) anatomic total shoulder arthroplasty.

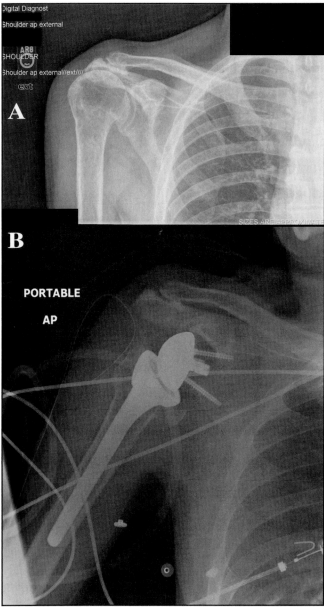

Figure 8-15. This patient with (A) GH arthritis and irreparable cuff insufficiency was treated with (B) reverse shoulder arthroplasty.

Conclusion

GH arthritis represents a continuum of diseases based on differential involvement of articular cartilage, bone, capsule, and rotator cuff. Clinical and radiographic features vary according to the type of arthritis present. The type and severity of arthritis present can be discerned on the basis of an accurate history, physical examination, radiographic evaluation, and CT scanning or MRI when indicated. When nonoperative treatment fails, surgical management is individualized by patient age, activity level, arthritis severity, rotator cuff integrity, bone deformity, and other factors. Shoulder replacement is indicated in many patients and provides predictable pain relief and improved function, especially when the rotator cuff is intact.

References

1. Edwards TB, Boulahia A, Kempf JF, Boileau P, Nemoz C, Walch G. The influence of rotator cuff disease on the results of shoulder arthroplasty for primary osteoarthritis: results of a multicenter study. *J Bone Joint Surg Am.* 2002;84A(12):2240-2248.
2. Iannotti JP, Norris TR. Influence of preoperative factors on outcome of shoulder arthroplasty for glenohumeral osteoarthritis. *J Bone Joint Surg Am.* 2003;85A(2):251-258.
3. Neer CS, Craig EV, Fukuda H. Cuff-tear arthropathy. *J Bone Joint Surg Am.* 1983;65(9):1232-1244.
4. McCoy SR, Warren RF, Bade HA III, Ranawat CS, Inglis AE. Total shoulder arthroplasty in rheumatoid arthritis. *J Arthroplasty.* 1989;4(2):105-113.
5. Neer CS. Replacement arthroplasty for glenohumeral osteoarthritis. *J Bone Joint Surg Am.* 1974;56(1):1-13.
6. Norris TR, Iannotti JP. Functional outcome after shoulder arthroplasty for primary osteoarthritis: a multicenter study. *J Shoulder Elbow Surg.* 2002;11(2):130-135.
7. Turkcapar N, Demir O, Atli T, et al. Late onset rheumatoid arthritis: clinical and laboratory comparisons with younger onset patients. *Arch Gerontol Geriatr.* 2006;42(2):225-231.
8. Cicuttini FM, Baker J, Hart DJ, Spector TD. Relation between Heberden's nodes and distal interphalangeal joint osteophytes and their role as markers of generalized disease. *Ann Rheum Dis.* 1998;57(4):246-248.
9. Hertel R, Ballmer FT, Lombert SM, Gerber C. Lag signs in the diagnosis of rotator cuff rupture. *J Shoulder Elbow Surg.* 1996;5(4):307-313.

10. Simovitch RW, Helmy N, Zumstein MA, Gerber C. Impact of fatty infiltration of the teres minor muscle on the outcome of reverse total shoulder arthroplasty. *J Bone Joint Surg Am.* 2007;89(5):934-939.
11. Lynch NM, Cofield RH, Silbert PL, Hermann RC. Neurologic complications after total shoulder arthroplasty. *J Shoulder Elbow Surg.* 1996;5(1):53-61.
12. Blazar PE, Williams GR, Iannotti JP. Spontaneous detachment of the deltoid muscle origin. *J Shoulder Elbow Surg.* 1998;7(4):389-392.
13. Edelson JG. Localized glenoid hypoplasia: an anatomic variation of possible clinical significance. *Clin Ortho Rel Res.* 1995;321:189-195.
14. Halverson PB, Cheung HS, McCarty DJ, Garancis J, Mandel N. "Milwaukee shoulder": association of microspheroids containing hydroxyapatite crystals, active collagenase, and neutral protease with rotator cuff defects: II. Synovial fluid studies. *Arthritis Rheum.* 1981;24(3):474-483.
15. Hamada K, Fukuda H, Mikasa M, Kobayashi Y. Roentgenographic findings in massive rotator cuff tears: a long-term observation. *Clin Ortho Rel Res.* 1990;254:92-96.
16. Walch G, Badet R, Boulahia A, Khoury A. Morphologic study of the glenoid in primary glenohumeral osteoarthritis. *J Arthroplasty.* 1999;14(6):756-760.
17. Cuomo F, Greller MJ, Zuckerman JD. The rheumatoid shoulder. *Rheum Dis Clin North Am.* 1998;57(4):246-248.

9

FRACTURES/TRAUMA

Michael R. Tracy, MD and Charles L. Getz, MD

INTRODUCTION

The shoulder girdle is composed of 3 bones (ie, scapula, clavicle, humerus), the joints linking these bones (ie, sternoclavicular [SC], acromioclavicular [AC], and glenohumeral [GH]), and the scapulothoracic articulation between the deep surface of the scapula and the posterior aspect of the thorax. High energy (eg, an automobile accident) or low energy (eg, a fall from standing height) trauma can injure any of the components of this complex.

Clavicle fractures are the most common shoulder fractures. These fractures account for 5% of all fractures seen in adults and have a bimodal age distribution peaking under the age

Cohen SB. *Musculoskeletal Examination of the Shoulder: Making the Complex Simple* (pp. 196-213). © 2011 SLACK Incorporated

of 40 years and over the age of 70 years.[1] Proximal humeral fractures account for 4% to 5% of all adult fractures,[2,3] while scapular fractures account for only 3% to 5% of fractures around the shoulder girdle and 1% of all adult fractures.[4,5]

This chapter will review fractures of the scapula, clavicle, and proximal humerus, as well as dislocations of the SC and GH joints (Chapter 4) and scapulothoracic dissociation. Injuries to the AC joint were addressed in Chapter 8.

HISTORY

For all patients with a traumatic mechanism of injury—but especially those who have been subjected to a high-energy trauma—standard primary and secondary trauma surveys should be completed. Once life-threatening injuries to the head, thorax, and abdomen have been either ruled out or identified and appropriately treated, the physician can focus on injuries to the shoulder girdle. While the majority of scapular fractures are seen in the setting of high-energy trauma, either high-energy or low-energy trauma (eg, fall from standing height) can result in significant injuries to the shoulder girdle.

A thorough history, including mechanism of injury and associated complaints, will provide invaluable information to the clinician. Patients often will be able to localize their pain to the clavicle or proximal humerus. Patients with scapular fractures may complain of deep pain in the shoulder blade. Patients who suffer a dislocation of the GH joint with spontaneous reduction will often relate a feeling of the joint slipping out of place. Description of sharp, shooting pain at the time of injury should raise suspicion for a brachial plexus injury, especially in patients who have experienced a hyperflexion and/or hyperabduction injury as the result of reaching out while falling. Patients who suffer a fracture without obvious traumatic mechanism (eg, while throwing a baseball) should be examined for a possible underlying bony lesion. In older patients who suffer injuries from a fall, the clinician should pursue a thorough investigation of how the injury happened (eg, mechanical fall by tripping versus syncope) to determine whether an underlying medical condition may have precipitated the injury. A history of previous falls and prior fragility fractures is also important.

EXAMINATION

Physical examination of the injured shoulder is similar to a basic musculoskeletal examination and should involve inspection, palpation, and neurovascular testing of the affected extremity (Table 9-1). Range of motion testing may be difficult after an acute injury. However, motor, sensory, and vascular examination both proximally at the shoulder and distally in the affected upper extremity are critical to detecting an underlying neurovascular injury. A significant number of patients with trauma to the shoulder girdle will demonstrate partial or complete neurologic injury, which is likely to recover without intervention.[6,7] Catastrophic injuries to the brachial plexus and its branches or to the axillary vasculature are relatively uncommon but can carry devastating consequences for the patient.[8]

To fully inspect the shoulder girdle, the patient's shoulders must be exposed to allow comparison to the contralateral side. Patients with an anterior SC dislocation may demonstrate a palpable, medial deformity. Patients with an anterior GH dislocation will have a prominent humeral head anteriorly and a void beneath the lateral deltoid.[1] Conversely, patients with a posterior GH dislocation will present with a prominent coracoid and a soft tissue void anteriorly.[9] Patients with fractures of the clavicle may or may not demonstrate obvious deformity depending on the degree of fracture displacement and the body habitus of the patient. Tenting of the skin by clavicular fracture fragments should be noted as this can lead to skin compromise.[1] The skin also should be examined carefully for small inside-out punctures indicative of an open fracture. Patients with proximal humeral fractures may not demonstrate any obvious deformity; however, local swelling and bruising will be observed. Palpation of the bruised or injured area will cause pain.

The neurovascular examination, including an assessment of the brachial plexus and its branches (Table 9-2), will be helpful in determining the urgency with which an injury to the shoulder girdle must be addressed. Function of the anterior and middle deltoid, innervated by the axillary nerve, can be assessed with minimal movement of the upper extremity.

Table 9-1

HELPFUL HINTS

Injury	Physical Examination	Imaging	Notes
Sterno-clavicular dislocation	Anterior: Palpable prominence Posterior: Difficulty with breathing, swallowing, and phonation	Serendipity view, CT scan	Vascular or cardiothoracic surgeon on stand-by for operative procedures
Clavicle fracture	Tenting of skin, puncture wounds signifying open fracture	Orthogonal views to assess displacement	Consider surgical intervention for shortened, displaced fractures in young, active patients
Gleno-humeral dislocation	Anterior: Loss of deltoid contour, humeral head palpable anteriorly Posterior: Arm locked in internal rotation	Shoulder trauma series including axillary view, subacute MRI as indicated	Reduce as quickly as possible
Proximal humerus fracture	Bruising and ecchymosis	Shoulder trauma series, CT scan, and angiography as indicated	Intervention dependent on fracture pattern and patient-related factors
Scapular injuries	Local tenderness to palpation	Shoulder trauma series, CT scan	Rule out other life-threatening injuries

Table 9-2

METHODS FOR EXAMINING THE TERMINAL BRANCHES OF THE BRACHIAL PLEXUS

Nerve	Motor	Sensory
Axillary	Deltoid	Lateral proximal humerus
Median	Index finger distal inter-phalangeal (DIP) flexion	Palmar surface index finger
Musculo-cutaneous	Biceps	Lateral forearm
Radial	Thumb extension	Dorsal first web space
Ulnar	Interossei	Palmar surface small finger

Sensory function of the axillary nerve is assessed at the lateral proximal humerus (ie, the sergeant's patch). It should be noted, however, that sensation in this area can be retained even when a significant deltoid motor deficit is present.[10] The musculocutaneous nerve provides motor function to the biceps muscle and sensation along the lateral aspect of the forearm via the lateral antebrachial cutaneous nerve. The radial nerve and its terminal motor branch, the posterior interosseous nerve, can be tested by resisting thumb extension and testing the dorsal surface of the first web space. The median nerve and its terminal motor branch, the anterior interosseous nerve, can be tested by resisting distal interphalangeal flexion of the index finger and testing the palmar surface of the index finger. Finally, the ulnar nerve can be tested by resisting finger abduction and adduction and testing the palmar surface of the small finger. Combined deficits suggest an injury proximal to the terminal branches of the brachial plexus. The presence of Horner's syndrome (ie, ptosis, miosis, and anhydrosis) is more consistent with a proximal supraclavicular injury than an infraclavicular injury.

Injury to the subclavian or brachial arteries, while uncommon, can be catastrophic. Furthermore, the high degree of collateral blood flow around the shoulder can mask even a major vascular insult. Capillary refill in the fingers of the affected extremity should be assessed. However, it is critical to palpate the radial and ulnar pulses and compare these to the contralateral side. A palpable difference in pulse should raise suspicion for a vascular injury. If there is a question, ankle:brachial index easily can be obtained in the office or the emergency room. Other signs suggestive of a vascular injury include an expanding mass, bruit, axillary hematoma, or excessive chest wall bruising.[8]

The final assessment is palpation and range of motion of the shoulder girdle. While gross deformity of the arm or clavicle will not require palpation for diagnosis, more subtle injuries can be discovered by manual palpation. When no fractures are present, gently ranging the shoulder may uncover a posterior dislocation if external rotation is severely restricted.

PATHOANATOMY

While bony or articular derangement of the shoulder girdle may be obvious and dramatic, the potential for injury to the surrounding soft tissue structures is often as important, if not more important. The brachial plexus and subclavian vessels course directly under the clavicle and can be injured by clavicular and proximal humeral fractures. The axillary nerve and brachial artery are susceptible to injury as they course past the GH joint. The trachea and esophagus rest posterior to the SC joint and can be compressed by a posterior SC dislocation.

The nature of the bony injury is influenced by the location of the fracture and the muscular attachments of the proximal and distal fragments. The sternocleiodmastoid, pectoralis major, and sternohyoid muscles all attach to the medial clavicle, while the lateral clavicle is anchored to the scapula by the coracoclavicular (CC) and AC ligaments. The anterior deltoid, trapezius, and clavicular head of the pectoralis major also attach to the lateral clavicle. The confluence of these muscular forces, along with the weight of the arm, lead to displacement and shortening of the clavicle after a mid-shaft fracture. This in turn disrupts the clavicle's function as a strut to maintain the appropriate length-tension relationship of its attached muscles.[1]

The proximal humerus will fracture along the lines of embryonic development. Neer[11] developed a classification scheme for proximal humerus fractures based on the patho-anatomy of the primary fracture fragments: the anatomic neck, the surgical neck, the greater tuberosity, and the lesser tuberosity. The supraspinatus, infraspinatus, and teres minor all insert on the greater tuberosity, while the subscapularis inserts on the lesser tuberosity. The pectoralis major and deltoid insert on the shaft of the humerus. These muscular attachments will determine the orientation of the fracture fragments, with separate fragments being defined by displacement greater than 1 cm or angulation greater than 45 degrees.

IMAGING

Standard radiographs of the shoulder include anteroposterior (AP), scapular Y, and axillary views. Obtaining an axillary view is critical to appropriately assessing the injured shoulder. This is the only view that can definitively rule out a GH dislocation. This is especially true in the case of posterior GH dislocations, as these injuries may be missed 60% to 79% of the time.[12] The axillary view is also the best view on which to see displacement of the greater and lesser tuberosities and splitting of the humeral head. If a standard axillary view cannot be obtained because of the patient's inability to abduct the arm secondary to pain, a Velpeau axillary can be obtained.[13]

For injuries to the other components of the shoulder girdle, specialized plain films may be helpful (see Table 9-1). Clavicle fractures should be imaged in at least 2 orthogonal planes; otherwise, displacement of the fracture may be underestimated. A 40-degree cephalic tilt serendipity view can help confirm a SC dislocation or injury.[1,14] A Zanca view provides excellent visualization of the AC joint by eliminating the scapula from the field.[14] A west point view will show the anterior glenoid rim and is useful for assessing the presence or absence of a bony Bankart lesion after acute GH dislocation. The Stryker notch view shows the posterosuperior humeral head and is useful for assessing the presence or absence of a Hill-Sachs lesion after acute GH dislocation.[15]

Advanced imaging of the shoulder girdle can include computed tomography (CT) scan, magnetic resonance imaging (MRI), and angiography. A CT scan may be helpful to better evaluate the position and bone quality of proximal humerus fractures, especially when portable trauma views of the shoulder provide inadequate information for diagnosis. CT scanning is also helpful when evaluating scapular fractures, including glenoid face and neck fractures, and scapular body fractures, which can be difficult to see on plain x-ray. While not necessary in the immediate postinjury period, an MRI may be obtained to assess the rotator cuff, labrum, and biceps tendon in patients who continue to have post-traumatic pain and weakness. There is a 15% incidence of rotator cuff tears in patients 40 years of age at the time of dislocation and a 40% incidence of rotator cuff tears in those 60 years of age at the time of dislocation.[16] MRI also may be necessary to evaluate the capsuloligamentous structures, labrum, and articular cartilage in younger patients who develop either gross or subtle instability after GH dislocation.[15] Finally, arteriography is crucial for any patient with a suspected vascular injury.[8]

TREATMENT

Most injuries to the shoulder girdle do not require emergency treatment. Open fractures, acute vascular compromise, and scapulothoracic disassociation do necessitate urgent operative intervention, and GH dislocations should be reduced as quickly as possible. However, the remainder of the injuries can be treated either conservatively or with staged operative intervention.

Sternoclavicular Dislocations

SC joint dislocations most commonly occur anteriorly. These injuries can be painful and present with a palpable anterior prominence. Closed reduction in the operating room can be attempted but often cannot be maintained. Chronic anterior SC dislocation is rarely symptomatic. For those dislocations that remain symptomatic, surgery can be performed to reconstruct the SC ligaments.[14,17-19]

Acute posterior SC dislocations are less common but are more likely to require acute intervention as they may compress vital neurovascular structures, including the innominate artery and vein, internal jugular vein, trachea, esophagus, vagus nerve, and phrenic nerve. Closed reduction is attempted in the operating room with the aid of a towel clip and a vascular or cardiothoracic surgeon on stand-by. Successful closed reductions will often be stable. In patients with irreducible dislocations or joint instability after reduction, reconstruction of the SC joint may be necessary. Again, this is done in conjunction with a vascular or cardiothoracic surgeon.[14,17,18]

Clavicle Fractures

The most common clavicle fractures occur in the mid-shaft (Figure 9-1A). The prevailing treatment modality had been nonoperative care with a sling or clavicle strap.[20] However, several recent studies have demonstrated that clavicular malunions can have significant effects on shoulder function and that malunions and nonunions after conservative treatment are much more common than originally thought.[1,21,22] As such, there has been a growing trend toward internal fixation of mid-shaft clavicle fractures, especially in young, active patients.[1] Operative treatment of mid-shaft clavicle fractures can be accomplished with either plating (Figure 9-1B) or intramedullary fixation (Figure 9-1C).

Fractures of the medial third of the clavicle are rare, and the current literature supports nonoperative treatment. Fractures of the lateral third of the clavicle (Figure 9-2) are less common than mid-shaft fractures but are more likely to progress to symptomatic nonunion or malunion when displaced. Nondisplaced fractures of the lateral third of the clavicle can be successfully treated with conservative measures; however, operative fixation is often needed for displaced lateral third fractures in young, active patients.[1]

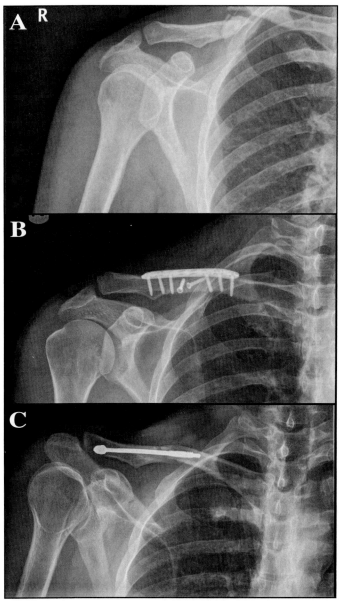

Figure 9-1. (A) AP x-ray of the clavicle showing a mid-shaft clavicle fracture. (B) Treatment of a mid-shaft clavicle fracture with a plate. (C) Treatment of a mid-shaft clavicle fracture with a clavicle pin.

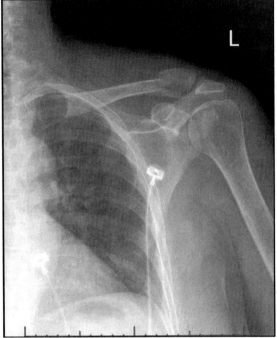

Figure 9-2. AP x-ray of a distal (lateral) third clavicle fracture.

Glenohumeral Dislocation

GH dislocations (see Chapter 4) most commonly occur anteriorly. Posterior dislocations often are seen in the setting of seizure or electrocution.[9] Patients who reach out during a fall and experience an upwardly directed force on a fully forward flexed arm can dislocate inferiorly, the so-called luxatio erecta. All shoulder dislocations should be reduced as quickly as possible.[15] Anterior and posterior dislocations can often be reduced in the emergency room with appropriate local anesthetic (eg, an intra-articular injection of lidocaine) and/or conscious sedation. Anterior and posterior dislocations that cannot be reduced under conscious sedation and inferior dislocation will require closed or open reduction under general anesthesia in an operating room.

If an anterior dislocation is successfully reduced in the emergency room, the standard of care has been a brief period of immobilization in a sling followed by early range of motion and strengthening[15]; however, immobilization in external rotation has recently been proposed to help prevent recurrent dislocations.[23] Evidence is also mounting that young, active first-time dislocators may have a better outcome with early operative stabilization.[24,25] Patients younger than 40 years at the time of initial dislocation are at increased risk for developing shoulder instability, with risk indirectly correlated with age at initial dislocation. Patients older than 40 are at high risk for a traumatic rotator cuff tear after shoulder dislocation and may require an MRI and appropriate surgical intervention.[15,26] Axillary nerve injuries may also be seen in this patient population.[26]

After reduction of a posterior GH dislocation, conservative therapy is appropriate for patients with a relatively small (ie, less than 25% of the articular surface) defect in the humeral head. However, patients with a defect greater than this may require surgical intervention to prevent instability.[9] Additionally, posterior GH dislocations often cause injury to the posterior capsulolabral structures, with or without an accompanying fracture of the glenoid rim. This injury, or posterior glenoid erosion caused by chronic posterior GH dislocation, may necessitate glenoid osteotomy or bone grafting to prevent recurrent dislocation.[9]

Proximal Humerus Fracture

Proximal humerus fractures (Figures 9-3 and 9-4) are classified by the number of displaced fracture fragments, with displacement greater than 1 cm or angulation greater than 45 degrees identifying separate fragments.[11] This classification scheme also helps to direct treatment options.[13] Additionally, the physiologic age of the patient, activity level, bone quality, and medical comorbidities are considered when determining the appropriate treatment.[27,28] The majority of proximal humerus fractures are nondisplaced and can be treated successfully with conservative management; however, approximately 15% of proximal humerus fractures are displaced and require surgical intervention.[3]

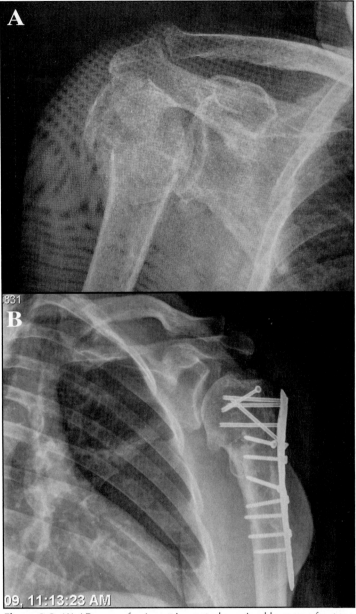

Figure 9-3. (A) AP x-ray of a 4-part impacted proximal humerus fracture. (B) A proximal humerus fracture treated with a plate-and-screw construct *(continued)*.

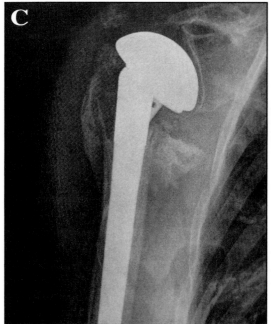

Figure 9-3 (continued). (C) A proximal humerus fracture treated with a hemiarthroplasty.

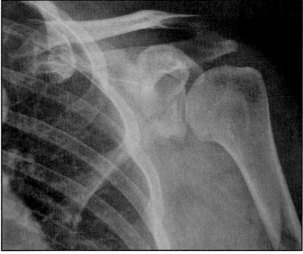

Figure 9-4. AP x-ray of a nondisplaced anterior glenoid fracture.

Two-part fractures at the surgical neck often can be treated conservatively with sling immobilization and early range of motion. If the fracture fragments are widely displaced, consideration may be given to closed reduction and percutaneous pinning or open reduction and plate fixation.[3,27] Two-part fractures involving either the greater tuberosity or the lesser tuberosity often are addressed surgically, as the pull of the rotator cuff muscles on these fragments can lead to further displacement and high rates of symptomatic malunion or nonunion.[27]

Three-part proximal humeral fractures often are treated with open reduction and internal fixation. In older, low-demand patients with badly comminuted fractures, some consideration may be given to hemi-arthroplasty.

Determining the optimal treatment for 4-part fractures can be difficult (see Figure 9-3A). In young patients with good bone stock, open reduction and internal fixation can be attempted, but functional results are mixed (see Figure 9-3B). Valgus-impacted 4-part fractures may also be addressed with surgical fixation as the blood supply to the humeral head often remains intact, resulting in a lower incidence of avascular necrosis.[3,29] Older, low-demand patients with poor bone quality are treated with hemi-arthroplasty (see Figure 9-3C). The key to successful treatment of these fractures with hemi-arthroplasty is reconstructing the original anatomy of the proximal humerus, especially the positioning of the greater and lesser tuberosities.[28] Recently, there has been research in treating low-demand patients with irreparable proximal humerus fractures with reverse total shoulder arthroplasty. Long-term results for this strategy are still not known.[30]

Scapular Fractures and Scapulothoracic Disassociation

Scapular body fractures are often treated nonoperatively, and nonunion is rare. However, these injuries often occur in the setting of high-energy trauma and are associated with other injuries.[4] A thorough examination of the patient with a scapular body fracture is necessary to rule out other, life-threatening injuries.

Fractures that extend into the scapular neck or glenoid fossa can be treated nonoperatively if they are not displaced significantly and if the shoulder girdle remains stable (see Figure 9-4). CT scanning will help determine the extent of the injury. Fractures of the glenoid neck with more than 1 cm of displacement or more than 40 degrees of angulation may benefit from surgical fixation.[4] Fractures of the glenoid fossa require surgical fixation if there is more than 5 mm of articular step-off or if there is subluxation of the humeral head.[4,31] The clavicle, scapular body, scapular spine, acromion, coracoid, AC ligaments, and CC ligaments form a superior suspensory complex around the shoulder joint. Injuries to a single element of this complex often are stable and can be treated conservatively. Double disruptions of the complex may lead to shoulder instability and may necessitate surgical intervention.[4,32,33] However, Williams and colleagues[34] have shown that ipsilateral fractures of the clavicle and glenoid neck are unstable only if the AC and CC ligaments are also injured.

Scapulothoracic dissociation is a rare but devastating injury. As the forequarter is traumatically separated from the axial skeleton, serious injury can occur to the neurovascular structures of the upper extremity. Urgent surgical intervention may be needed to save the limb, though prognosis is poor.[4] Functional outcome is directly correlated with the severity of the neurologic injury.[35]

CONCLUSION

Injuries to the shoulder girdle are common. They can include fractures of the scapula, clavicle, or humerus; dislocations of the SC and GH joints; and scapulothoracic dissociation. Thorough history and physical examination and appropriate imaging are necessary to accurately assess the bony injury and any associated soft tissue injuries. Treatment can range from conservative care to nonemergent surgical intervention to urgent surgical intervention.

REFERENCES

1. Kim W, McKee MD. Management of acute clavicle fractures. *Orthop Clin N Am*. 2008;39:491-505.
2. Buhr AJ, Cooke AM. Fracture patterns. *Lancet*. 1959;1:531-536.
3. Magovern B, Ramsey ML. Percutaneous fixation of proximal humerus fractures. *Orthop Clin N Am*. 2008;39:405-416.
4. Lapner PC, Uhthoff HK, Papp S. Scapula fractures. *Orthop Clin N Am*. 2008;39:459-474.
5. Rowe CR. Fractures of the scapula. *Surg Clin North Am*. 1963;43:1565-1571.
6. Blom S, Dahlback LO. Nerve injuries in dislocations of the shoulder joint and fractures of the neck of the humerus: a clinical and electromyographical study. *Acta Chir Scand*. 1970;136(6):461-466.
7. de Laat EA, Visser CP, Coene LN, et al. Nerve lesions in primary shoulder dislocations and humeral neck fractures: a prospective clinical and EMG study. *J Bone Joint Surg Br*. 1994;76(3):381-383.
8. Zarkadas PC, Throckmorton TW, Steinmann SP. Neurovascular injuries in shoulder trauma. *Orthop Clin N Am*. 2008;39:483-490.
9. Kowalsky MS, Levine WN. Traumatic posterior glenohumeral dislocation: classification, pathoanatomy, diagnosis, and treatment. *Orthop Clin N Am*. 2008;39:519-533.
10. Steinmann SP, Moran EA. Axillary nerve injury: diagnosis and treatment. *J Am Acad Orthop Surg*. 2001;9(5):328-335.
11. Neer CS. Displaced proximal humeral fractures. Part I: classification and evaluation. *J Bone Joint Surg Am*. 1970;52:1077-1089.
12. Matsen FA III, Titelman RM, Lippitt SB, et al. Glenohumeral instability. In: Rockwood CA Jr, Matsen FA III, Wirth MA, et al, eds. *The Shoulder* Vol. 2. 3rd ed. Philadelphia, PA: Saunders; 2004:655-794.
13. Robinson BC, Athwal GS, Sanchez-Sotelo J, Rispoli DM. Classification and imaging of proximal humerus fractures. *Orthop Clin N Am*. 2008;39:393-403.
14. MacDonald PB, Lapointe P. Acromioclavicular and sternoclavicular joint injuries. *Orthop Clin N Am*. 2008;39:535-545.
15. Dodson CC, Cordasco FA. Anterior glenohumeral joint dislocations. *Orthop Clin N Am*. 2008;39:507-518.
16. Henry JH, Genung JA. Natural history of glenohumeral dislocation: revisited. *Am J Sports Med*. 1982;10:135-137.
17. Bicos J, Nicholson GP. Treatment and results of sternoclavicular joint injuries. *Clin Sports Med*. 2003;22(2):359-370.
18. Iannotti J, Williams Jr G. *Disorders of the Shoulder: Diagnosis and Management*. 2nd ed. Philadelphia, PA: Lippincott Williams & Wilkins; 2007:979-1006.
19. Nettles JL, Linscheid RL. Sternoclavicular dislocations. *J Trauma*. 1968;8(2):158-164.
20. Neer CS. Nonunion of the clavicle. *JAMA*. 1960;172:1006-1011.

21. Ledger M, Leeks N, Ackland T, et al. Short malunions of the clavicle: an anatomic and functional study. *J Shoulder Elbow Surg.* 2005;14:349-354.

22. McKee MD, Pedersen EM, Jones C, et al. Deficits following non-operative treatment of displaced, mid-shaft clavicle fractures. *J Bone Joint Surg Am.* 2006;88:35-40.

23. Itoi E, Hatakeyama Y, Kido T, et al. A new method of immobilization after traumatic anterior dislocation of the shoulder: a preliminary study. *J Shoulder Elbow Surg.* 2003;12:413-415.

24. Bottoni CR, Wilckens JH, DeBerardino TM, et al. A prospective, randomized, evaluation of arthroscopic stabilization versus nonoperative treatment of acute, first-time shoulder dislocations. *Am J Sports Med.* 2002;30(4):576-580.

25. Kirkley A, Griffin S, Richards C, et al. Prospective randomized clinical trial comparing the effectiveness of immediate arthroscopic stabilization versus immobilization and rehabilitation in first traumatic anterior dislocations of the shoulder. *Arthroscopy.* 1999;15:507-514.

26. Neviaser RJ, Neviaser TJ, Neviaser JS. Concurrent rupture of the rotator cuff and anterior dislocation of the shoulder in the older patient. *J Bone Joint Surg Am.* 1988;70(9):1308-1311.

27. Drosdowech DS, Faber KJ, Athwal GS. Open reduction and internal fixation of proximal humerus fractures. *Orthop Clin N Am.* 2008;39:429-439.

28. Krishnan SG, Bennion PW, Reineck JR, Burkhead WZ. Hemiarthroplasty for proximal humeral fracture: restoration of the gothic arch. *Orthop Clin N Am.* 2008;39:441-450.

29. Jakob RP, Miniaci A, Anson PS, et al. Four-part valgus impacted fractures of the proximal humerus. *J Bone Joint Surg Br.* 1991;73:295-298.

30. Martin TG, Iannotti JP. Reverse total shoulder arthroplasty for acute fractures and failed management after proximal humeral fractures. *Orthop Clin N Am.* 2008;39:451-457.

31. Goss TP. Fractures of the glenoid cavity. *J Bone Joint Surg Am.* 1992;74(2):299-305.

32. Goss TP. Scapular fractures and dislocations: diagnosis and treatment. *J Am Acad Orthop Surg.* 1995;3(1):22-33.

33. Goss TP. Double disruptions of the superior shoulder suspensory complex. *J Orthop Trauma.* 1993;7(2):99-106.

34. Williams GR Jr, Naranja J, Klimkiewicz J, et al. The floating shoulder: a biomechanical basis for classification and management. *J Bone Joint Surg Am.* 2001;83-A(8):1182-1187.

35. Zelle BA, Pape HC, Gerich TG, et al. Functional outcome following scapulothoracic dissociation. *J Bone Joint Surg Am.* 2004;86-A(1):2-8.

FINANCIAL DISCLOSURES

Dr. Geoffrey S. Baer has not disclosed any relevant financial relationships.

Dr. Michael G. Ciccotti has no financial or proprietary interest in the materials presented herein.

Dr. Steven B. Cohen is a consultant for Smith & Nephew Endoscopy and Knee Creations.

Dr. David R. Diduch has no financial or proprietary interest in the materials presented herein.

Dr. Charles L. Getz has no financial or proprietary interest in the materials presented herein.

Dr. George Paul Hobbs has no financial or proprietary interest in the materials presented herein.

Dr. Gregg J. Jarit has no financial or proprietary interest in the materials presented herein.

Dr. Scott Montgomery has no financial or proprietary interest in the materials presented herein.

Dr. William B. Morrison is on the medical advisory board for GE Medical Systems and is a consultant for Apriomed.

Dr. Mark W. Rodosky has no financial or proprietary interest in the materials presented herein.

Dr. James R. Romanowski has no financial or proprietary interest in the materials presented herein.

Dr. Michael Shin has no financial or proprietary interest in the materials presented herein.

Dr. Misty Suri is on the speaker's bureau for Arthrex and is a consultant for Breg and Arthrex.

Dr. Michael R. Tracy has no financial or proprietary interest in the materials presented herein.

Dr. Gerald R. Williams Jr is a consultant for Depuy, Depuy Mitek, and Tornier. He receives royalties from Depuy for shoulder arthroplasty and text book royalties from Lippincott. He also has stock in Invivo Therapeutics.

INDEX

Wait...There's More!

Throughout the *Musculoskeletal Examination Series*, you will find a thorough review of the most common pathologic conditions, techniques for diagnosis, and appropriate treatment methods. These pocket-sized books include very clear photographic demonstrations, tables, and charts, taking complex subjects and bringing them to a level that will be welcomed by all.

Series Editor: Steven B. Cohen, MD

Musculoskeletal Examination of the Foot and Ankle: Making the Complex Simple

Shepard R. Hurwitz, MD; Selene Parekh, MD

275 pp., Soft Cover, 2011, ISBN 13 978-1-55642-919-4, Order #19193, **$44.95**

Musculoskeletal Examination of the Hip and Knee: Making the Complex Simple

Anil Ranawat, MD; Bryan T. Kelly, MD

480 pp., Soft Cover, 2011, ISBN 13 978-1-55642-920-0, Order #19207, **$48.95**

Musculoskeletal Examination of the Shoulder: Making the Complex Simple

Steven B. Cohen, MD

240 pp., Soft Cover, 2011, ISBN 13 978-1-55642-912-5, Order #19126, **$44.95**

Musculoskeletal Examination of the Spine: Making the Complex Simple

Jeffrey A. Rihn, MD; Eric B. Harris, MD

275 pp., Soft Cover, 2011, ISBN 13 978-1-55642-996-5, Order #19965, **$44.95**

Musculoskeletal Examination of the Elbow, Wrist, and Hand: Making the Complex Simple

Randall Culp, MD

275 pp., Soft Cover, 2011, ISBN 13 978-1-55642-918-7, Order #19185, **$44.95**

Please visit **www.slackbooks.com** to order any of the above title

24 Hours a Day...7 Days a Week!